LECTURE NOTES ON

Infectious Diseases

BIBHAT K. MANDAL
FRCP
Honorary Consultant Physician in Infectious Diseases
North Manchester General Hospital
Director, International Collaborating Centre for Research in Tropical Medicine, Calcutta, India

EDMUND G.L. WILKINS
FRCP FRCPath
Consultant Physician

EDWARD M. DUNBAR
FRCP
Consultant Physician

both of the Department of Infectious Diseases
and Tropical Medicine
North Manchester General Hospital
Manchester

RICHARD T. MAYON-WHITE
FRCP FFPHM
Consultant Public Health Physician
Department of Public Health
The John Radcliffe Hospital
Oxford

Sixth edition

Blackwell
Publishing

© 1969, 1974, 1980, 1984, 1996, 2004 by Blackwell Publishing Ltd
Blackwell Publishing, Inc., 350 Main Street, Malden, Massachusetts 02148-5020, USA
Blackwell Publishing Ltd, 9600 Garsington Road, Oxford OX4 2DQ, UK
Blackwell Publishing Asia Pty Ltd, 550 Swanston Street, Carlton, Victoria 3053, Australia

First published 1969
Second edition 1974
Revised reprint 1975
Third edition 1980
Fourth edition 1984
Fifth edition 1996
Reprinted 1999
Reprinted 2001
Sixth edition 2004

Library of Congress Cataloging-in-Publication Data

Lecture notes on infectious diseases / Bibhat K. Mandal . . . [et al.].—6th ed.
          p. ;   cm.
Includes index.
    ISBN 1-4051-0820-7
1. Communicable diseases.
    [DNLM: 1. Infections. 2. Communicable Diseases.   WC 100 L4713 2003]   1. Mandal,
Bibhat K.
RC111.L36 2003
616.9—dc22   2003018220

ISBN 1-4051-0820-7

A catalogue record for this title is available from the British Library

Set in 9/11.5 Gill Sans by SNP Best-set Typesetter Ltd., Hong Kong
Printed and bound in the United Kingdom by MPG Books Ltd, Bodmin, Cornwall

Commissioning Editor: Vicki Noyes
Editorial Assistant: Nic Ulyatt
Production Editor: Karen Moore
Production Controller: Kate Charman

For further information on Blackwell Publishing, visit our website:
http://www.blackwellpublishing.com

Contents

Section 4: Infection in Special Groups, Zoonoses, Tropical Diseases and Helminths

Preface

The scale of changes seen in the field of infectious diseases over the past few years has necessitated this new edition. We have retained the previous edition's 'user-friendly' format of presentation that proved popular with readers. However, many chapters have been completely or substantially rewritten, others have been fully updated. New features include a separate chapter on infection problems of special groups like travellers, injecting drug users and others, sepsis and septic shock, infections with bioterrorism potential and information on global occurrence in the epidemiology sections wherever relevant. Useful websites have been listed for up-to-date information on fast-moving topics like AIDS, bioterrorism, travel-related infections and SARS.

Some expansion of the text has been unavoidable but this has been partly compensated for by sacrificing the colour pictures of the previous editions, which did not contribute much to the knowledge because of their small numbers. The *Lecture Notes on Infectious Diseases* thus remains a modest-sized book. It aims to provide a concise framework of knowledge on the diagnosis and management of major infectious diseases in a practical, 'easy-to-read' format. We hope the book will continue to prove useful to medical students, young doctors in training, practising physicians and other health-care providers engaged in infection management.

B.K. Mandal
E.G.L. Wilkins
E.M. Dunbar
R.T. Mayon-White

List of Abbreviations

ADEM	acute disseminated encephalomyelitis	DOT	directly observed therapy
AFB	acid-fast bacilli	DSS	dengue shock syndrome
ALT	alanine aminotransferase	EAggEC	enteroaggregative *Escherichia coli*
ARDS	adult respiratory distress syndrome	EBL	European bat lyssavirus
ARV	antiretroviral	EBV	Epstein–Barr virus
ASD	atrial septal defect	ECG	electrocardiogram
ASO	antistreptolysin O	ECM	erythema chronicum migrans
AST	aspartate aminotransferase	EIA	enzyme immunoassay
BAL	bronchoalveolar lavage	EIEC	enteroinvasive *Escherichia coli*
BCG	bacille Calmette–Guérin	ELISA	enzyme-linked immunoadsorbent assay
BMT	bone marrow transplant	EM	electron micrograph
BSE	bovine spongiform encephalopathy	EPEC	enteropathogenic (enteroadherent) *Escherichia coli*
CAH	chronic active hepatitis	EPI	expanded programme of immunization
CBD	common bile duct		
CCHF	Crimean–Congo haemorrhagic fever	ERCP	endoscopic retrograde cholangiopancreatography
CDC	Centers for Disease Control and Prevention, Atlanta, USA	ESBLs	extended-spectrum β-lactamases
CI	cellular immunity	ESR	erythrocyte sedimentation rate
CIE	countercurrent immunoelectrophoresis	ET	exfoliative toxin
		ETEC	enterotoxigenic *Escherichia coli*
CIN	cervical intraepithelial neoplasia	EV71	enterovirus 71
		FBCt	full blood count
CJD	Creutzfeldt–Jakob disease	FTA	fluorescent treponemal antibody
CMI	cell-mediated immunity		
CMV	cytomegalovirus	GBS	group B streptococci
CPH	chronic persistent hepatitis	GBS	Guillain–Barré syndrome
CRP	C-reactive protein	GI	gastrointestinal
CSF	cerebrospinal fluid	gp	glycoprotein
CT	computed tomography	GUM	genitourinary medicine
CXR	chest X-ray	HAART	highly active antiretroviral therapy
DEET	diethyltoluamide		
DHF	dengue haemorrhagic fever	HACEK	*Haemophilus, Actinobacillus, Cardiobacterium, Eikenella* and *Kingella*
DIC	disseminated intravascular coagulopathy		

HAV	hepatitis A virus	MR	magnetic resonance
HBV	hepatitis B virus	MRSA	methicillin-resistant
HCV	hepatitis C virus		*Staphylococcus aureus*
HDV	hepatitis delta virus	MSU	midstream urine
HEV	hepatitis E virus	MTCT	mother-to-child transmission
HFRS	haemorrhagic fever and renal	NA	nucleic acid
	syndrome	NE	nephropathica epidemica
HFV	hepatitis F virus	NGU	non-gonococcal urethritis
HHV	human herpesvirus	NLV	Norwalk-like virus
Hib	*Haemophilus influenzae* type b	NNRTIs	non-nucleoside reverse
HIV	human immunodeficiency virus		transcriptase inhibitors
HIVAN	HIV-associated nephropathy	NRTIs	nucleoside reverse
HPIV	human parainfluenza virus		transcriptase inhibitors
HPS	hantavirus pulmonary	NSAID	non-steroidal anti-inflammatory
	syndrome		drug
HPV	human papillomavirus	OCP	ova, cyst and parasite
HTLV	human T-cell lymphotropic virus	OHL	oral hairy leucoplakia
HUS	haemolytic–uraemic syndrome	OI	opportunistic infection
HZ	herpes zoster	OMPs	outer membrane proteins
ICP	intracranial pressure	OPV	oral polio vaccine
ICU	intensive care unit	ORS	oral rehydrating solution
IDU	injection drug user	PAN	polyarteritis nodosa
IFAT	antibody titre by	PCNSL	primary central nervous system
	immunofluorescence		lymphoma
IgM	immunoglobulin type M	PCP	*Pneumocystis carinii* pneumonia
IM	intramuscular	PCR	polymerase chain reaction
IP	incubation period	PDA	patent ductus arteriosus
IPV	inactivated poliovirus vaccine	PEP	postexposure prophylaxis
IRIS	immune reconstitution	PET	positron emission tomography
	syndrome	PGL	persistent generalized
ITP	idiopathic thrombocytopenic		lymphadenopathy
	purpura	PHN	postherpetic neuralgia
IV	intravenous	PID	pelvic inflammatory disease
IVDU	intravenous drug user	PIs	protease inhibitors
JE	Japanese B encephalitis	PMC	pseudomembranous colitis
LCM	lymphocytic choriomeningitis	PMFL	progressive multifocal
LDH	lactate dehydrogenase		leucoencephalopathy
LFTs	liver function tests	PUO	pyrexia of unknown origin
LGV	lymphogranulovenereum	RFLP	restriction fragment length
LP	lumbar puncture		polymorphism
MAC	*Mycobacterium avium*	RIG	rabies immunoglobulin
	complex	RMAT	rapid microagglutination test
MBCs	minimum bactericidal	RMSF	Rocky Mountain spotted
	concentrations		fever
MI	myocardial infarction	RPR	rapid plasma reagin (test)
MICs	minimum inhibitory	RSV	respiratory syncytial virus
	concentrations	RT	reverse transcriptase
MMR	measles, mumps and rubella	RTI	respiratory tract infection

SARS	severe acute respiratory distress syndrome	TSST	toxic shock syndrome toxin
		USS	ultrasound scan
Sd I	*Shigella dysenteriae* type I	UTI	urinary tract infection
SIRS	systemic inflammatory response syndrome	vCJD	variant Creutzfeldt–Jakob disease
SLE	systemic lupus erythematosus	VDRL	Venereal Disease Research Laboratory
SRSV	small round structured virus		
STARI	Southern tick-associated rash illness	VHF	viral haemorrhagic fever
		VL	viral load
STD	sexually transmitted disease	VSD	ventricular septal defect
STI	sexually transmitted infection	VTEC	verocytotoxin-producing
TB	tuberculosis		*Escherichia coli*
TBE	tick-borne encephalitis	WCCt	white cell count
TEN	toxic epidermal necrolysis	WHO	World Health Organization
TPHA	*Treponema pallidum* haemagglutination assay	ZDV	Zidovudine
		ZN	Ziehl–Neelsen
TSS	toxic shock syndrome		

Genus Abbreviations

Genus	Species
Actinobacillus	A. actinomycetemcomitans
Ancylostoma	A. braziliensis, A. duodenale
Ascaris	A. lumbricoides
Aspergillus	A. flavus, A. fumigatus, A. niger
Bacillus	B. anthracis
Bacteroides	B. fragilis
Bartonella	B. bacilliformis, B. elizabethae, B. henselae, B. quintana
Bordetella	B. parapertussis, B. pertussis
Borrelia	B. burgdorferi, B. lonestari
Brucella	B. abortus, B. canis, B. melitensis, B. suis
Campylobacter	C. fetus, C. jejuni
Candida	C. albicans, C. glabrata, C. parapsilosis, C. tropicalis
Capnocytophaga	C. canimorsus
Chlamydia	C. trachomatis
Clostridium	C. botulinum, C. difficile, C. histolyticum, C. novyii, C. perfringens, C. septicum, C. tetani, C. welchii
Coccidioides	C. immitis
Corynebacterium	C. diphtheriae, C. haemolyticum, C. ulcerans
Coxiella	C. burnetii
Cryptococcus	C. neoformans
Cryptosporidium	C. parvum
Cyclospora	C. cayetanensis
Diphylobothrium	D. latum
Eikenella	E. corrodens
Encephalocytozoon	E. hellem, E. intestinali
Entamoeba	E. histolytica
Enterobius	E. vermicularis
Enterococcus	E. faecalis, E. faecium
Enterocytozoon	E. bieneusi
Erysipelothrix	E. rhusiopathiae
Escherichia	E. coli
Franciscella	F. tularensis
Fusobacterium	F. necrophorum
Giardia	G. lamblia (intestinalis)
Haemophilus	H. ducreyi, H. influenzae type b
Helicobacter	H. pylori

Genus	Species
Histoplasma	H. capsulatum
Klebsiella	K. pneumoniae
Legionella	L. pneumophila
Leishmania	L. aethiopica, L. braziliensis, L. chagasi, L. donovani, L. infantum, L. major, L. mexicana, L. peruviana, L. tropica
Leptospira	L. australis, L. autumnalis, L. balam, L. biflexa, L. canicola, L. grippotyphosa, L. hardjo, L. icterohaemorrhagiae, (L. icterohaemorrhagica), L. interrogans, L. pomona
Listeria	L. monocytogenes
Loa	L. loa
Madurella	M. mycetomatis
Malassezia	M. furfur
Mycobacterium	M. africanum, M. avium complex (M. scrofulaceum, M. intracellulare and M. avium), M. avium-intracellulare, M. bovis, M. catarrhalis, M. chelonei, M. fortuitum, M. genavensae, M. haemophilum, M. kansasii, M. leprae, M. malmoense, M. marinum, M. tuberculosis, M. ulcerans, M. xenopi
Mycoplasma	M. gentilium, M. pneumoniae
Necator	N. americanus
Neisseria	N. gonorrhoeae, N. meningitidis
Onchocerca	O. volvulus
Penicillium	P. marneffeii
Paragonimus	P. westermani
Pasteurella	P. multocida
Pediculus	P. humanus capitis, P. humanus corporis
Plasmodium	P. falciparum, P. malariae, P. ovale, P. vivax
Pneumocystis	P. carinii
Proteus	P. mirabilis
Pseudomonas	P. aeruginosa, P. pseudomallei
Pthirus	P. pubis
Rickettsia	R. akari, R. conori, R. prowazeki, R. rickettsii, R. tsutsugamushi, R. typhi
Salmonella	S. agona, S. cholerae-suis, S. dublin, S. enteritidis, S. hadar, S. heidelberg, S. indiana, S. paratyphi, S. typhi, S. typhimurium, S. virchow
Sarcoptes	S. scabiei
Schistosoma	S. mansoni, S. japonicum, S. haematobium
Shigella	S. boydii, S. dysenteriae, S. flexneri, S. sonnei
Spirillium	S. minor
Staphylococcus	S. aureus, S. epidermidis, S. saprophyticus
Streptobacillus	S. moniliformis
Streptococcus	S. bovis, S. intermedius, S. milleri, S. mitior, S. mutans, S. pneumoniae, S. pyogenes, S. salivarius, S. sanguis, S. viridans
Strongyloides	S. stercoralis
Taenia	T. saginata, T. solium
Toxocara	T. canis, T. catis
Toxoplasma	T. gondii
Treponema	T. carateum, T. pallidum, T. pallidum var. endemic syphilis, T. pertenue

Genus	Species
Trichinella	T. spiralis
Trichomonas	T. vaginalis
Trichuris	T. trichura
Trypanosoma	T. brucei gambiense, T. brucei rhodesiense, T. cruzi
Tunga	T. penetrans
Ureaplasma	U. urealyticum
Vibrio	V. cholerae, V. parahaemolyticus
Wuchereria	W. bancrofti
Xanthomonas	X. maltophilia
Yersinia	Y. enterocolitica, Y. pestis

General Topics

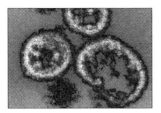

Introduction

Changing pattern of infectious diseases

The incidence of infection is an ever-changing pattern, which is one reason why the study of infectious diseases is so interesting. While better sanitation, clean food and water supplies, good housing, personal hygiene, vaccines and drugs have controlled some diseases, new diseases emerge and other diseases are newly recognized to be infections. In the resource-poor developing countries infectious diseases continue to cause significant morbidity and mortality. In the developed countries, the picture of infectious disease at the beginning of the 21st century can be summarized as follows.

• There is one disease, smallpox, that has been eradicated from the world, although there is a potential threat of its deliberate release

• A second disease, poliomyelitis, is close to global eradication, having been absent from the Americas and Europe for over 3 years

• Infections which have virtually disappeared as endemic disease: cholera, typhus, diphtheria

• Infections which have become much less common or less virulent: measles, mumps, rubella, whooping cough, tetanus, tuberculosis, *Haemophilus influenzae* type b diseases, scarlet fever

• Infections whose incidence has remained un-changed are: respiratory infections, chickenpox (except in countries practising universal child-hood varicella vaccination, e.g. the USA) and herpes zoster, infantile gastroenteritis (much less severe), infections of the nervous system (except *Haemophilus* meningitis), neonatal infections, urinary infections

• Infections which have increased are: sexually transmitted infections, infections in immuno-compromised, debilitated and intensive care unit patients, methicillin-resistant *Staphylococcus aureus* (MRSA) outbreaks of infection in hospitals, infections in intravenous drug users

• Infections that increased in between 1980 and 2000, but now show signs of being better con-trolled are: salmonella and listeria infections associated with food, *Clostridium difficile*

• Infection associated with the increasing travel to tropical countries for business and holidays are: malaria, enteric fever, amoebiasis, helminthiasis, exotic viral infections, traveller's diarrhoea

• New infection problems are: human im-munodeficiency virus (HIV) infection, variant Creutzfeldt–Jakob disease, multidrug resis-tance in pneumococci, salmonellae, tuber-culosis and staphylococci, severe acute respiratory distress syndrome (SARS).

In the last decade (up to 2003), five global factors have emerged as forces that could cause further changes:

3

1 Climate change, and specifically global warming, could extend the geographical range of infections like malaria.

2 Population increase, accompanied by environmental degradation, could result in inadequate supplies of safe food and water.

3 Large numbers of people migrating for safety or economic and social reasons may bring high rates of diseases like tuberculosis into cities in both the developing and the developed countries.

4 Xenotransplantation and genetic modification could, in theory, result in new human pathogens, despite the safeguards taken to prevent this.

5 Bioterrorism and other deliberate releases of biological agents may be perpetrated to extort money (see Table 1.1).

The major aetiological agents of infectious diseases identified between 1972 and 2003 are listed in Table 1.2.

Transmission of infection

Infection spreads by one of the following methods.

Airborne

Infection is exhaled from the case or carrier by coughing, sneezing or speaking, in invisible respiratory droplets of moisture which are inhaled by the new host. The microorganisms may adhere to dust or textiles, leaving infected dust which may still transmit infection. Skin scales are an important source of contaminated dust. Dust may be carried by air currents, but rarely for distances of more than a few metres.

Diseases spread by airborne routes include:

Exanthemata: measles, rubella, chickenpox, scarlet fever.

Mouth and throat infections: diphtheria, tonsillitis, mumps, herpes stomatitis.

Respiratory tract infections: whooping cough, influenza and other respiratory virus infections, pulmonary tuberculosis.

General: meningococcal and staphylococcal infection.

Intestinal

Infection present in the bowel excreta of a case or carrier is ingested by a fresh host. Transmission may be immediate and direct via infected fingers, eating utensils, clothing, toilets, etc., or indirect via food or water.

Diseases spread by the intestinal route include typhoid and paratyphoid, salmonellosis, dysentery, cholera, gastroenteritis, poliomyelitis and other enterovirus infections, and viral hepatitis A and E.

In another group of ingestion diseases, transmission is direct from contaminated food. This group includes brucellosis, Q fever,

Table 1.1 Infections that might be released deliberately as acts of terrorism.

Infection	Notes
Anthrax	Tested as a biological weapon, 1939–1944; thought to be stocked as a weapon, 1992; used in USA, 2001 (see p. 109)
Botulism	Risk of food or water contamination (see p. 85)
Plague	(see p. 110)
Smallpox	Considered as a biological weapon 1950–1989 (see p. 103)
Tularaemia	(see p. 111)
Viral haemorrhagic fevers	(see p. 228)

Sources for more information
www.hpa.org.uk/infections/topics_az/deliberate_release/menu.htm
www.who.int/csr/delibepidemics/en/
www.bt.cdc.gov/

Year	Agent	Disease
1972	Norwalk-like viruses	Diarrhoea outbreaks
1973	Rotaviruses	Major cause of infantile diarrhoea worldwide
1975	Astroviruses	Diarrhoea (outbreaks)
1975	Parvovirus B19	Fifth disease and aplastic crises in chronic haemolytic anaemia
1976	*Cryptosporidium parvum*	Diarrhoea
1977	Ebola virus	Ebola haemorrhagic fever
1977	*Legionella pneumophila*	Legionnaires' disease
1977	Hantaan virus	Haemorrhagic fever with renal syndrome
1977	*Campylobacter* spp.	Diarrhoea
1980	Human T-cell lymphotropic virus 1 (HTLV-1)	Adult T-cell leukaemia/lymphoma; tropical spastic paresis
1982	HTLV-2	Hairy cell leukaemia
1982	*Borrelia burgdorferi*	Lyme disease
1983	Human immunodeficiency viruses (HIV-1, HIV-2)	Acquired immunodeficiency syndrome and related illnesses
1983	*Escherichia coli* O157	Haemorrhagic colitis and haemolytic–uraemic syndrome
1983	*Helicobacter pylori*	Gastritis and gastric ulcers
1988	Human herpesvirus 6 (HHV-6)	Exanthema subitum (roseola infantum)
1989	*Erlichia* spp.	Human erlichiosis
1989	Hepatitis C virus (HCV)	Parenterally transmitted non-A non-B hepatitis
1990	Human herpesvirus 7 (HHV-7)	Pityriasis rosea
1990	Hepatitis E virus (HEV)	Enterically transmitted non-A non-B hepatitis
1991	Hepatitis F (HFV)	Severe non-A non-B hepatitis
1992	*Bartonella henselae*	Cat-scratch disease
1993	Sin nombre virus	Hantavirus pulmonary syndrome
1994	Sabia virus	Brazilian haemorrhagic fever
1994	Human herpesvirus 8 (HHV-8)	Kaposi's sarcoma
1995	Hendravirus	Meningitis, encephalitis
1996	Prion disease related to bovine spongiform encephalopathy	Variant Creutzfeldt–Jakob disease (vCJD)
1997	Enterovirus 71 (EV71)	Epidemic encephalitis
1998	Nipahvirus	Meningitis, encephalitis
1999	West Nile virus	Encephalitis (New York)
2003	New Coronavirus	Severe acute respiratory syndrome (SARS)

Table 1.2 Major aetiological agents of infectious diseases identified between 1972 and 2003.

salmonellosis, trichinellosis and other helminth infections.

Direct contact

Infection may be transmitted directly by local skin contact. This mostly involves cutaneous infections and includes impetigo and scabies.

Venereal route

Infection may be transmitted by sexual contact, including syphilis, gonorrhoea, lymphogranuloma venereum and herpes genitalis infection, HIV and hepatitis B infection.

Insect or animal bite

Infections transmitted by bites include malaria, leishmaniasis, trypanosomiasis, typhus, rabies and simian herpesvirus infection.

Blood-borne

Some infections are commonly transmitted via infected blood or blood products, e.g. hepatitis B, HIV, hepatitis C.

These do not cover all the complex routes by which disease spreads. For example, leptospirae excreted in rats' urine may contaminate stagnant water and later penetrate the intact skin of a human host bathing in the water, or tetanus spores from the faeces of herbivorous animals may contaminate pasture land and years later may enter a wound and cause human disease.

Other diseases may spread by two or more alternative routes. For example, tuberculosis commonly spreads by airborne infection, but may spread via milk by ingestion or even by direct skin contact.

Control measures

Notification

All countries have some system of reporting select infections for public health purposes: 'notification'. The list of diseases, the mechanism of reporting and the legal basis for the notifications vary from country to country. But they have in common that the public interest justifies giving

Table 1.3 Notifiable diseases.

medical information about a patient to a public health service, that there are control measures that can be taken to prevent further spread or that the disease is important enough to require surveillance. In the UK, the list of notifiable diseases specified by the Public Health (Infectious Diseases) Regulations 1988 is detailed in Table 1.3, and the notification is made to the Consultant in Communicable Disease Control (CCDC) for the district in which the patient is living.

Collecting data from clinical microbiologists on laboratory-proven infections also provides surveillance in most countries that have well-established microbiological services. Some countries have sentinel or 'spotter' general practitioners to consistently and quickly report conditions like influenza-like illnesses.

Communicable disease control

Outbreaks and serious cases of infectious disease are investigated by the public health specialists (CCDC and community infection control nurses in the UK) and environmental officers, in collaboration with microbiologists and infectious disease consultants. The source of infection, mode of spread, contacts and occupational circumstances are all investigated and appropriate measures carried out, including the isolation and treatment of patients and the immunization and control of carriers and contacts.

Acute encephalitis	Paratyphoid fever
Acute poliomyelitis	Plague
Anthrax	Rabies
Cholera	Relapsing fever
Diphtheria	Rubella
Dysentery (amoebic or bacillary)	Scarlet fever
Food poisoning	Smallpox
Leprosy	Tetanus
Leptospirosis	Tuberculosis
Malaria	Typhoid fever
Measles	Typhus
Meningitis	Viral haemorrhagic fevers
Meningococcal septicaemia (without meningitis)	Viral hepatitis
Mumps	Whooping cough
Ophthalmia neonatorum	Yellow fever

International control measures

In cases of smallpox, cholera, plague and yellow fever (officially referred to as 'diseases subject to the regulations') the World Health Organization arranges an interchange of information to enable the necessary public health and preventive measures to be carried out. The World Health Organization also regularly exchanges information on a further group of infections which are kept under surveillance, and this includes poliomyelitis, epidemic influenza, louse-borne relapsing fever and louse-borne typhus fever.

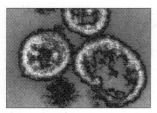

Host Response to Infection and Immunization

Defence against infection

The human body is continually exposed to a wide range of potentially pathogenic microbial organisms in its environment as well as within itself, yet most people do not experience recurrent or continued infections. This is due to the existence of a complex set of defence mechanisms.

Prevention of entry

The *skin* and *epithelia* of the gastrointestinal (GI), respiratory and genitourinary systems are effective barriers to microbial invasion. These barriers are supported by secreted *mucus* and *ciliary movement* of the respiratory tract, and by *gastric acidity* and *commensal flora*.

Non-immune defence mechanisms

These become active very rapidly to eliminate microbes that manage to enter the body and are active against any invader.

• *Complement system:* this involves a group of plasma proteins which undergo a cascade of activation when triggered by an antigen (i.e. a microbe). The resultant production of active components help bacteriolysis and the antimicrobial activities of neutrophils and macrophages.

• *Neutrophils:* these migrate to the site of invasion and destroy microbes by phagocytosis.

• *Mononuclear cells/macrophages:* these are also able to ingest bacteria that have been coated with complement components or antibody.

• *Natural killer lymphocytes:* these can kill target cells without requiring antigenic or antibody stimulation and appear to help in resisting virus infections.

• *Interferons:* these are a group of glycoproteins which help to prevent viral replication in host cells.

• *Inflammation* at the site of invasion is produced by the above factors and helps to contain further spread of infection.

Immune defence mechanisms

These are formed in response to specific microbes to produce immunity against that microbe and take several days or longer to develop. There are two types: humoral and cellular.

Humoral immunity

Humoral immunity is characterized by production of antibodies by specific clones of B lymphocytes, bearing receptors for the particular invading microbe, triggered into proliferation by T-helper lymphocytes.

• *Antibodies* are immunoglobulins which combat infection in a number of ways: activation of

complement, helping phagocytosis, neutralization of microbes
• Immunoglobulin type M (IgM) antibody is produced initially (disappearing after several months), followed by a more pronounced and persistent production of IgG antibodies
• IgA antibodies are secreted on mucosal surfaces. These are important first-line defences against future infection (coating microbes and preventing adherence to mucosal cells)
• At the same time, greater numbers of *memory cells* (antigen-specific B cells) are produced, so future exposure provokes a quicker and more vigorous antibody response
• Humoral immunity is responsible for recovery from many acute infections and prevention of further attacks in future.

Cellular immunity
• T lymphocytes are the primary effectors of cellular immunity
• Activated T cells produce cytotoxic cells which destroy infected cells, and cytokines which prevent microbes from replicating within cells
• CD4-bearing T cells (T-helper cells) also produce lymphokines which activate cytotoxic T-killer cells and produce inflammation which are characteristics of delayed hypersensitivity reaction
• Cellular immunity is important for recovery from virus infections, many fungal (e.g. candida, histoplasma, cryptococcus, toxoplasma, cryptosporidium) and helminthic (e.g. strongyloides) infections and some bacterial (e.g. mycobacteria) infections.

Sepsis and septic shock

Sepsis is present when a patient with infection exhibits certain systemic manifestations of inflammatory response, i.e. fever or hypothermia, tachycardia, and leucocytosis or leucopenia (systemic inflammatory response syndrome, SIRS).
• 'Severe sepsis' is characterized by the additional presence of multiorgan dysfunction

• When hypotension unresponsive to fluid resuscitation supervenes the patient has 'septic shock'
• 'Septicaemia' is diagnosed when bacteraemia is associated with SIRS.

Epidemiology
Severe sepsis and septic shock are the leading causes of death in intensive care units (ICUs) worldwide and account for 2–11% of all hospital or ICU admissions in the USA and Europe.
    The incidence has increased in recent decades because of the growth in:
• intensive care management
• immunosuppressed population
• elderly population
• population living longer with chronic diseases
• intravenous drug misuse
• microbial resistance.

Microbial aetiology and source
• Most cases of severe sepsis are caused by the following organisms in about equal proportions:
    • Gram-negative bacilli (*Escherichia coli*— commonest, *Klebsiella*, *Pseudomonas aeruginosa*) from urinary tract, lung, abdomen
    • Gram-positive cocci (mainly staphylococci and streptococci) from skin and soft tissue, intravenous devices, lung
• Fungi, mostly *Candida* (GI tract, long venous lines), account for about 5% of cases
• Meningococci are an important cause of community-acquired septic shock
• Unusual organisms: *Capnocytophaga* (dog bite), babesiosis, Rocky Mountain spotted fever (RMSF).

Pathogenesis
Inflammatory response at the site of infection, a result of the host's non-specific and specific immune mechanisms, defends it against the invading microbe by containing its growth and eventually destroying it.
    If the microbe manages to overcome this local defence and escapes into the surrounding tissues or bloodstream, a complex cascade of interactions may be triggered which involves

microbial factors (toxins, cell wall components) and host factors (complement pathways, leucocytes and humoral mediators, e.g. cytokines) and results in coagulation abnormalities, tissue injury, vascular collapse and multiorgan dysfunction.

Clinical features

The following are commonly present in severe sepsis:

• fever and tachycardia
• hyperventilation
• liver, lung and renal dysfunction
• hypotension
• encephalopathy, usually resulting from poor perfusion rather than due to tissue damage
• rash: in meningococcaemia, toxic shock syndrome, *Capnocytophaga* infection, ecthyma gangrenosum, RMSF.

Laboratory findings

• Leucocytosis or leucopenia
• Thrombocytopenia
• Disseminated intravascular coagulopathy— microangiopathic red cells, raised D dimers and prolonged prothrombin time
• Raised urea, creatinine, bilirubin, transaminases, lactate
• Respiratory alkalosis, later metabolic acidosis
• Chest X-ray—changes of acute respiratory distress syndrome

Investigations

• Search for microbes in blood, urine, catheter tip, inflamed sites
• X-ray and scans to localize site of infection
• Monitoring organ functions.

Management

• Empirical antibiotic therapy (see Table 3.1)
• Haemodynamic and nutritional support
• Organ support
• Adjunctive treatments: many such treatments have undergone clinical trials aimed at neutralization of toxins, minimization of inflammatory response (low-dose corticosteroids), neutralization of proinflammatory cytokines (anti-TNF), correction of coagulation abnor-

malities (activated protein C) but at best gave 'mixed' results and none is in routine use.

Prognosis

• Between 30 and 50% die
• Fatal outcome is related more to factors such as advanced age, neutropenia and pre-existing diseases than to specific infections.

Immunization

Immunization is the induction of artificial immunity by giving preformed antibodies as immunoglobulin (passive immunization) or by giving an antigen as a vaccine (active immunization). The term *vaccination* is interchangeable with *active immunization*.

• Immunoglobulin used for passive immunization is either non-specific, collected from pooled human blood donations (e.g. normal immunoglobulin for protection against hepatitis A), or specific, formed from high-titre sera of humans (e.g. zoster immunoglobulin to prevent chickenpox). The protection given by immunoglobulin is short-lived (because of protein catabolism), varying from 1 to 4 months, according to the dose of immunoglobulin, the level of antibodies in it and the disease in question.

• Active immunization uses antigens which can be killed organisms (e.g. whole-cell pertussis vaccine), live organisms with low (attenuated) virulence (e.g. measles vaccine), inactivated bacterial products (e.g. diphtheria toxoid), or selected antigens of the particular organism (e.g. pneumococcal capsular polysaccharide). The antigens can be made by fraction of disrupted pathogens, or by genetic engineering.

Routine immunization of children

All countries have vaccination programmes for young children. These programmes differ because of variations in the incidence of preventable infections and in the health priorities. The World Health Organization's expanded programme of immunization (EPI) and the globalization of drug manufacture are reducing this variation. In all countries the key objective of childhood vaccination programmes is to vaccinate as many children as possible. Vaccines that are used in childhood programmes are listed in Table 2.1.

In some other countries, injected inactivated polio vaccines are preferred to the live oral vaccine, hepatitis B vaccine is given in the first year of life, bacille Calmette–Guérin (BCG) is given to all infants, and measles, mumps and rubella (MMR) vaccine is boosted later in childhood.

Timing of vaccinations

- Vaccinations at birth are for infections that become an immediate risk: congenital hepatitis B when the mother is infected, and BCG in places of high tuberculosis prevalence
- Primary courses normally start at 2 months of age when the immune system starts to mature
- Killed organism vaccines generally require multiple injections to produce sustained immunity (e.g. diphtheria and tetanus toxoid and pertussis vaccines)

Table 2.1 Vaccines used in routine childhood programmes.

- Boosters are needed either to give lifelong immunity, or to reinforce protection when facing a high risk of infection, e.g. tetanus vaccine after an injury
- Live vaccines and conjugate polysaccharide vaccines given after the first birthday typically need only a single injection, but schedules of second dose of MMR vaccine are used in North America, Europe and Australia to raise herd immunity to these infections
- Live vaccines should be given together or spaced at 2–3-week intervals to avoid interference from the immune response to one or other injection
- BCG should also be spaced 3 or more weeks from other vaccinations because of temporary immunomodulation.

Vaccinations for travel, work and other purposes

Whereas the vaccination of children should be a routine and universal procedure, vaccinations for travel and occupational risks should be selected. Recommendations vary according to the perceptions of, and attitudes to, risk and the balance of benefits and costs (in which inconvenience is highly rated). Table 2.2 lists the vaccines that are available, with some possible uses.

Age	Vaccine
At birth	Bacille Calmette–Guérin (BCG)†, hepatitis B†
Between 2 and 6 months capsular polysaccharide	Diphtheria, tetanus and pertussis vaccine* Conjugated *Haemophilus influenzae* vaccine* Polio vaccine* Conjugate meningococcal C vaccine* Conjugate pneumococcal vaccine† Hepatitis B vaccine
Second year	Measles, mumps and rubella (MMR) vaccine* Varicella vaccine
Between 18 months and school entry	Boosters of diphtheria, tetanus, pertussis, polio and MMR vaccines*
Teenage	Boosters of diphtheria booster, tetanus and polio vaccine* Hepatitis B vaccine

Those used in the UK in 2002 are marked * if universal and † if selective.

Table 2.2 Vaccinations for travel and other purposes.

Vaccine	Nature	Indications
Anthrax	Cell-free culture filtrate	Workers exposed to imported animal hides and bones, people at risk of germ warfare
Cholera	Whole cell, killed	Poor efficacy of short duration; no longer recommended
Diphtheria	Low-dose toxoid	Travellers to countries where diphtheria is not rare, and contacts of cases
Hepatitis A	Inactivated virus	Frequent travellers to countries where hepatitis is common
Hepatitis B	Recombinant surface antigen	Most health-care workers, people going to live in high endemic countries, injecting drug users, sexual partners of cases and carriers, patients at genitourinary clinics
Influenza A and B	Inactive viral components	People who are likely to have complications from influenza because of old age, diseases of the heart, lung or kidneys, immunosuppression or diabetes
Japanese B encephalitis	Formalin-inactive virus	Travellers who stay for more than 1 month during the rainy season in rural areas of Asia where this disease occurs in epidemics (usually areas where rice growing and pig farming coexist)
Meningococcal A, C, W135 and Y	Capsular polysaccharide	Travellers to places with epidemic group A or W135 meningococcal meningitis and for people in local outbreaks
Pneumococcal	Capsular polysaccharide	People with an increased risk of severe pneumococcal disease because of asplenia, diabetes, immunosuppression or diseases of the heart, lung or kidneys
Polio	Live attenuated (oral) or inactivated (injected)	Boosters for travellers to countries where polio is still present and for contacts during outbreak
Rabies	Inactivated cell-culture-derived virus	People who have been bitten by rabid or possibly rabid animals and people who have an increased risk of such exposure through work with animals or travellers to remote areas in endemic countries
Smallpox	Vaccinia (live virus)	People who may be exposed to smallpox in laboratories, clinical work or military duties. Ring vaccination of contacts of a case. Very rarely for mass immunization
Tetanus	Formalin-inactivated toxin (toxoid)	Following tetanus-prone injury if unimmunized or reinforcing dose more than 10 years previously
Tick-borne encephalitis	Inactivated virus	Camping or walking in forests of north and central Europe in late spring/summer
Typhoid	Oral attenuated (live) or purified Vi antigen	Travellers to places where typhoid is endemic
Yellow fever	Live attenuated virus	Travellers to African and South American countries where yellow fever virus exists

Contraindications to immunization

There are few absolute, and many false, contraindications to immunization.

• The one important contraindication to vaccination is a severe adverse reaction to an earlier dose: anaphylaxis, encephalitis or severe local inflammation

• Rubella vaccine is contraindicated in pregnancy although vaccination has been found not to cause fetal damage. Yellow fever or polio vaccines should not be withheld when there is risk of significant exposure

• A period of acute infection is not the best time to give a vaccine but the minor symptoms of common illnesses of childhood should not delay immunization. Antibiotics are not a contraindication

• Live vaccines should not be administered to patients who are immunocompromised. However, MMR appears to be safe for human immunodeficiency virus (HIV)-positive children whereas inactivated polio should be used instead of oral live vaccine

• Asthma, eczema and stable neurological diseases are not contraindications. Egg allergy is a contraindication for influenza vaccine, and previous anaphylactic reaction to egg contraindicates MMR, influenza and yellow fever vaccines.

Chemotherapy

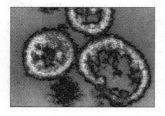

Antibiotics are the second most commonly used class of drug. One in three hospital patients receives antimicrobials, accounting for 25% of total drug costs. One in 20 develops adverse reactions, occasionally severe. Their rational use is therefore important and requires knowledge of the infective causes of presenting syndromes, the spectrum of activity of antimicrobials, the principles of pharmacokinetics, the contraindications and interactions of drugs, and where to obtain assistance in making the choice (*British National Formulary*).

Inappropriate prescribing

Above all, a doctor must be able to prescribe safely. Familiarity with a small list of antimicrobials is to be encouraged. In many situations antibiotics are not necessary, or are used inappropriately. Instances include:
- tonsillopharyngitis (usually viral)
- gastroenteritis (usually self-limiting if bacterial)
- wound infection (physical cleansing more important)
- where culture of an organism reflects contamination or colonization but not infection (lack of knowledge of significant pathogens)
- multidose surgical prophylaxis (often continued beyond 24 h).

Besides exposing the patient to the risk of un-necessary side-effects, inappropriate antibiotic use will add to the selection pressure for resistant strains and may delay the start of the correct treatment. It may also lead to *Clostridium difficile*-associated colitis.

Undertreatment is another problem. The most frequent errors of prescription are choosing the wrong drug and mistakes with antibiotic dose, duration or route of administration.

Antibiotic selection and use

Range of antibiotics

A wide selection of antibiotics is available, and it is important to become familiar with the major representatives from each class.

Beta-lactams
- Inhibit peptidoglycan cell wall synthesis
- Synergism with aminoglycosides
- Penicillin allergy in 1% (anaphylaxis 0.01%). 10–15% of those with penicillin allergy also react to cephalosporins
- Benzylpenicillin (G) and penicillin V have a narrow spectrum of activity (streptococci, *Neisseria*, clostridia); ampicillin and amoxicillin extend this to include *Listeria*, *Haemophilus influenzae* and enterococci; amoxicillin–clavulanic acid (co-amoxiclav) extends this further to cover *Staphylococcus aureus*, anaerobes and many 'coliforms'

- Flucloxacillin (penicillinase-resistant) is narrow spectrum and first line for *Staphylococcus aureus*
- Piperacillin (ureidopenicillin) and ticarcillin (carboxypenicillin) are broad-spectrum penicillins active against enterococci, anaerobes, 'coliforms' and *Pseudomonas aeruginosa*. They are often combined with a β-lactamase inhibitor
- Early cephalosporins have modest Gram-positive and weak Gram-negative activity (e.g. cefalexin)
- Cefuroxime, cefotaxime and ceftriaxone have a broader spectrum including *S. aureus*, streptococci, *Neisseria*, 'coliforms' and *H. influenzae*
- Ceftazidime has excellent Gram-negative spectrum including *P. aeruginosa*
- Cephalosporins do not have activity against enterococci
- Aztreonam has good Gram-negative activity but is inactive against Gram-positive organisms
- Imipenem and meropenem (carbapenems) have excellent Gram-negative, Gram-positive and anaerobic activity.

Aminoglycosides

- Bind to 30S subunit of ribosome, inhibiting protein synthesis
- Major drugs: gentamicin, tobramycin, amikacin and netilmicin
- Excellent Gram-negative spectrum including *P. aeruginosa*
- Useful where significant resistant to quinolones and β-lactams
- Synergistic with penicillin against enterococci and streptococci, and ampicillin against listeria
- No anaerobic activity
- Post-antibiotic effect allows once-daily administration
- Essential to monitor drug levels
- Toxic to cochlea and kidney (greater in the elderly).

Quinolones

- Inhibit DNA replication by interference with enzymes involved in coiling
- Major drugs: ciprofloxacin, ofloxacin, norfloxacin, levofloxacin

- Excellent Gram-negative spectrum; limited Gram-positive activity
- Newer quinolones (e.g. moxifloxacin) have improved Gram-positive action
- High tissue and intracellular levels
- Ciprofloxacin good for *P. aeruginosa*, enteric pathogens and atypical pneumonias but increasing resistance amongst many organisms reflecting widespread use.

Macrolides

- Bind to 50S subunit of ribosome, inhibiting protein synthesis
- Major drugs: erythromycin, azithromycin and clarithromycin
- Have good Gram-positive action and some anaerobic and enterococcal activity
- Useful for penicillin-allergic patients and atypical pneumonia
- Azithromycin particularly useful for single dose treatment of *Chlamydia trachomatis*.

Tetracyclines

- Prevent tRNA binding to ribosomes, inhibiting protein synthesis
- Major drugs: tetracycline, oxytetracycline, minocycline and doxycycline
- Modest broad spectrum but use limited to specific conditions (acne, *Chlamydia*, brucellosis, typhus, malaria, Lyme disease, syphilis etc.)
- Contraindicated in children and pregnant women.

Miscellaneous

Sulphonamides and trimethoprim
- Inhibit folate synthesis from para-aminobenzoic acid
- Major drugs: sulfamethoxazole with trimethoprim (co-trimoxazole), sulphadoxine with pyrimethamine (Fansidar) and sulfadiazine
- Inactive against anaerobes
- Co-trimoxazole useful as a second-line agent for chest and urinary tract infections (UTIs) and is the drug of choice for *Pneumocystis carinii*
- Fansidar is used for malaria treatment and prophylaxis
- Trimethoprim has modest broad-spectrum activity, useful only for UTIs.

Glycopeptides
- Inhibit peptidoglycan cell wall synthesis
- Major drugs: vancomycin and teicoplanin
- Excellent Gram-positive spectrum; inactive against Gram-negative organisms
- First-line drug for MRSA and many strains of enterococci and *Staphylococcus epidermidis*
- Essential to monitor drug levels because of renal toxicity.

Chloramphenicol
- Prevents tRNA binding to ribosome, inhibiting protein synthesis
- Good cerebrospinal fluid (CSF) penetration
- Broad spectrum including anaerobes
- Dose dependent and idiosyncratic bone marrow suppression.

Rifampicin
- Inhibits RNA synthesis
- First-line drug for tuberculosis
- Excellent Gram-positive spectrum; limited Gram-negative activity (*Legionella*, *Neisseria meningitidis*).

Nitroimidazoles
- Inhibit DNA replication
- Major drugs: metronidazole, tinidazole
- Excellent anaerobic and good protozoal spectrum.

Lincosamides
- Bind to 50S subunit of ribosome, inhibiting protein synthesis
- Major drugs: clindamycin, lincomycin
- Good Gram-positive activity; clindamycin has additional anaerobic activity
- Risk of *C. difficile* infection.

Fusidic acid
- Inhibits protein synthesis
- Narrow spectrum; additional drug for *S. aureus* infection.

Linezolid
- Oxazolidone with activity against Gram-positive bacteria

- Most importantly, active against methicillin-resistant *Staphylococcus aureus* (MRSA)
- IV and oral formulations.

Quinupristin/dalfopristin
- Bind to 50S subunit of ribosome, inhibiting protein synthesis
- Streptogrammin active against MRSA and highly resistant *E. faecium*
- Only available in intravenous formulation.

Choosing an antibiotic

In choosing an antibiotic, the doctor must:
- make a diagnosis as to the type of infection (e.g. pneumonia)
- have knowledge of both the organisms implicated and suitable antibiotics with activity against these organisms. The importance of whether an antibiotic is bactericidal or bacteriostatic is theoretical
- make a choice for the individual taking patient factors into account (e.g. pregnancy, age, renal or liver compromise, vomiting, allergy) and drug factors (e.g. penetration to site of disease, drug interactions)
- make use of the results of cultures and antibiotic susceptibility testing when they become available which will allow rationalization of the empirical therapy. Occasionally, the *in vivo* responses do not correspond, such as with aminoglycoside sensitivity with *Salmonella* (*in vitro* sensitive, *in vivo* resistant)
- take into account the need to monitor certain drugs where the toxic–therapeutic ratio is small (e.g. gentamicin, vancomycin).

The dose and route of administration must then be decided and an assessment as to the duration of total therapy made.
- Dose will depend upon weight, age, renal function, severity of infection and toxicity of the agent (e.g. cefuroxime is preferred to gentamicin for severe Gram-negative infections in the elderly with impaired renal function)
- Route of administration depends mainly on the severity and site of infection; the presence of malabsorption, venous access and likely compliance may also be determining factors. If similar

serum and tissue levels can be achieved by using high-dose oral antibiotic, there is no special need to continue intravenous (IV) antibiotics
• Duration of therapy rests on the response to therapy and likelihood of relapse (e.g. tuberculosis) or failure (e.g. endocarditis) with inadequate treatment.

Combinations

In certain situations, combination therapy is indicated.
• To reduce the likelihood of drug resistance (e.g. tuberculosis)
• To provide synergy (e.g. benzylpenicillin and gentamicin for treating Streptococcus viridans and enterococcal endocarditis)
• Where there is known or likely mixed infection (e.g. cefuroxime and metronidazole for gastrointestinal or biliary tract infections)
• To treat infection empirically where the likely organism is unknown (e.g. septic shock, neonatal sepsis, severe community-acquired pneumonia). In practice, this is the commonest reason
• Where clinical studies have shown that survival is improved (e.g. piperacillin and gentamicin in Gram-negative infections in immuno-compromised hosts)
• To cover the possibility of initial antibiotic resistance (e.g. vancomycin and cefotaxime where multiresistant S. pneumoniae is endemic).

Other forms of combination therapy include:
• a β-lactam antibiotic and β-lactamase inhibitor (e.g. amoxicillin and clavulanic acid)
• imipenem and cilastatin (a renal enzyme inhibitor reducing renal metabolism)
• probenecid with penicillins, reducing renal and CSF excretion, thereby increasing CSF and serum levels.

Combination therapy must be balanced by the potential disadvantages, namely interaction before administration (e.g. chloramphenicol and erythromycin), differential pharmacokinetics and tissue penetration which may select resistance, additional cost and combined adverse effects.

Chemoprophylaxis

Prophylaxis can be broadly divided into primary (preventing initial infection or disease) and secondary (preventing recurrent disease), although these terms are used loosely. Examples of prophylaxis in medicine include:
• penicillin for preventing recurrent attacks of rheumatic fever
• amoxicillin (or other) for preventing endocarditis during dental and other operative procedures in patients with valvular lesions
• co-trimoxazole for preventing Pneumocystis carinii pneumonia in human immunodeficiency virus (HIV) patients with CD4 counts of $< 200 \times 10^9$ cells/mm³
• aciclovir for preventing herpes simplex and ganciclovir for preventing cytomegalovirus (CMV) infections after bone marrow transplantation
• rifampicin and isoniazid for preventing subsequent relapse of a subclinical primary tuberculous infection
• rifampicin for preventing spread of N. meningitidis and H. influenzae nasopharyngeal carriage from a close contact of the index case
• erythromycin for preventing Bordetella pertussis infection in unvaccinated close home contacts.

Surgical antibiotic prophylaxis is mainly primary and should be given to the following categories of patients:
• those undergoing 'clean-contaminated' or 'contaminated' surgery (e.g. most gastrointestinal, biliary and gynaecological surgery)
• those receiving a prosthetic implant (e.g. orthopaedic and vascular)
• those with reduced resistance to infection
• where there is a break in aseptic technique intraoperatively.

The purpose of surgical prophylaxis is to cover the operative period when the site may become contaminated: antibiotics should not be given for more than 24 h.

The problems of resistance

There has been increasing resistance against many antibiotics.

• S. aureus: in resource-rich nations, approximately half of bacteraemic isolates are MRSA. In the UK, the major type is EMRSA-16, which has additional resistance to gentamicin and ciprofloxacin. Strains with reduced susceptibility to vancomycin have been reported

• S. pneumoniae: throughout the world, resistance is increasing. Types of resistance vary (see Chapter 5). Many countries in Europe report much higher levels (e.g. Spain) than the UK where 7% of strains show penicillin resistance and 15% erythromycin resistance

• Group A and B β-haemolytic streptococci: erythromycin resistance is increasing (5% in UK). No penicillin resistance in Group A β-haemolytic streptococci has ever been reported.

• Enterococci: vancomycin resistance is increasing, particularly in Enterococcus faecium where it has reached 27% in the UK.

• Enterobacteriaceae: increasing levels of resistance to broad-spectrum β-lactams (through extended-spectrum β-lactamases, ESBLs), ciprofloxacin and gentamicin are being seen

• Campylobacter: one-fifth of isolates are now resistant to ciprofloxacin and 1–5% to erythromycin

• Salmonella: multiresistant non-typhi and typhi isolates are increasingly being recovered.

There are many microbial methods of resistance:

• modification of binding proteins (β-lactams)
• reduced permeability (aminoglycosides)
• production of enzymes (β-lactams)
• modification of target (quinolones)
• active efflux (tetracyclines).

Antibiotic policy

The aim of an antibiotic policy is to provide a guide for the initial antibiotic therapy of a clinical syndrome. Supplementary to this is the achievement of an effective yet economic policy with reduced likelihood of encouraging the development of resistant organisms. Once the diagnosis has been confidently established and especially where an organism has been recovered, the antibiotic therapy can be modified accordingly. To be effective, the policy must be:

• straightforward, sensible and applicable to the types of infection seen within the hospital
• cost conscious (but not cost driven) and advise antibiotics that are routinely stocked by the pharmacy and available in the emergency drugs cupboard
• well publicized, audited at regular intervals and safe.

To formulate the policy, a committee should be formed including a microbiologist, a pharmacist and a physician. This group would decide on clinical infections to be included, look at the organisms implicated, examine the local resistance patterns and choose the optimal antibiotic from an appropriate shortlist. Several factors have to be taken into account when writing the policy (e.g. acceptable failure rate, cost, available formulations), and thought must also be given to the age group of the hospital population, the likelihood of the infection being nosocomial, and the possibility of concomitant renal or liver compromise. The policy must be well publicized and supported by all grades of staff to be successful.

Empirical antibiotics for specific infections

Table 3.1 gives advice on the choice of antibiotic(s) when the microbiological cause is not known. The route of administration is IV unless stated otherwise. There are many alternative choices to the ones given.

Table 3.1 Empirical antibiotics for specific infections.

Infection/age of patient	Causes	Recommended antibiotics
Central nervous system infections		
Meningitis		
<1 month	'Coliforms', group B β-haemolytic Streptococcus, Listeria	Cefotaxime and ampicillin
Community acquired	As above but also P. aeruginosa	Ceftazidime and ampicillin
Hospital acquired	N. meningitidis, H. influenzae (type b), S. pneumoniae	Cefotaxime
1 month–5 years	N. meningitidis, S. pneumoniae	Cefotaxime (benzylpenicillin if rash present)
Age > 5–50 years	S. pneumoniae, Gram-negative bacilli, Listeria	Cefotaxime and ampicillin
Age > 50 years	S. epidermidis, proprionibacterium, S. aureus, 'coliforms', P. aeruginosa	Vancomycin and ceftazidime (± intraventricular gentamicin with vancomycin)
Intraventricular shunt		
Cerebral abscess	Streptococcus milleri, other streptococci, 'coliforms', anaerobes, P. aeruginosa (otitic source), S. aureus (systemic)	Benzylpenicillin, metronidazole and cefotaxime (ceftazidime or ciprofloxacin if otitic source)
Encephalitis	Herpes simplex, enteroviruses, M. pneumoniae, influenza	Aciclovir and clarithromycin
Extradural spinal abscess	S. aureus, 'coliforms'	Flucloxacillin and cefotaxime
Bone and joint infections		
Osteomyelitis		
Age < 5 years	S. aureus, Streptococcus pyogenes, H. influenzae (type b)	Flucloxacillin and cefotaxime
Age > 5 years	S. aureus	Flucloxacillin and fucidin (or rifampicin or gentamicin)
Vertebral	S. aureus, 'coliforms'	Flucloxacillin and cefotaxime
Post metal implant	S. aureus, S. epidermidis	Vancomycin

Continued on p. 20.

Table 3.1 (continued)

Infection/age of patient	Causes	Recommended antibiotics
Septic arthritis		
Age < 5 years	*S. aureus, S. pyogenes, H. influenzae* (type b)	Flucloxacillin and cefotaxime
Age > 5 years	*S. aureus, S. pyogenes,* group B β-haemolytic *Streptococcus, N. gonorrhoeae,* 'coliforms'	Flucloxacillin and cefotaxime
Prosthetic joint	*S. aureus, S. epidermidis*	Vancomycin
Gastrointestinal infections		
Cholecystitis, cholangitis or diverticulitis	Anaerobes, 'coliforms', enterococci	Cefuroxime and metronidazole (or co-amoxiclav; or gentamicin, ampicillin and metronidazole)
Liver, pelvic or intra-abdominal abscess	*S. milleri,* other streptococci, enterococci, 'coliforms', *S. aureus*	Co-amoxiclav; or gentamicin, ampicillin and metronidazole
Diarrhoea or dysentery	*Campylobacter, Salmonella, Shigella, E. coli* O157, *C. difficile*	Ciprofloxacin (antibiotics not indicated if *E. coli* O157) Metronidazole (*C. difficile*)
Genitourinary tract infections		
Cystitis	'Coliforms', enterococci, *P. aeruginosa, Staphylococcus saprophyticus, S. aureus*	Trimethoprim, co-amoxiclav (oral), ciprofloxacin
Pyelonephritis	As above	Cefuroxime
Perinephric abscess	'Coliforms', *S. aureus*	Cefotaxime and flucloxacillin
Pelvic inflammatory disease	*N. gonorrhoeae, C. trachomatis,* 'coliforms', anaerobes, enterococci	Co-amoxiclav (oral) and doxycycline or metronidazole and doxycycline
Respiratory tract infections		
Pneumonia		
Community acquired	*S. pneumoniae, S. aureus, H. influenzae, Coxiella, Mycoplasma, Legionella, Chlamydia psittaci, C. pneumoniae*	Clarithromycin and cefotaxime

Hospital acquired	'Coliforms', P. aeruginosa, S. aureus (± MRSA), Enterobacter, Acinetobacter, S. pneumoniae, Legionella	Ceftazidime (or meropenem) and teicoplanin
Aspiration	Anaerobes, streptococci, S. aureus and 'coliforms'	Cefuroxime and metronidazole (or co-amoxiclav)
Infective exacerbation of chronic lung disease	S. pneumoniae, H. influenzae, moraxella catarrhalis	Clarithromycin and cefotaxime (moderate/severe) Co-amoxiclav, or co-trimoxazole (mild)

Ear, nose, throat and eye infections

Otitis media, sinusitis		
Acute	S. pneumoniae, S. pyogenes, H. influenzae	Amoxicillin, or co-trimoxazole or co-amoxiclav
Chronic	As above, also anaerobes, 'coliforms'	Co-amoxiclav
Otitis externa	S. aureus P. aeruginosa	Flucloxacillin (oral) Topical agents
Tonsillitis	S. pyogenes	Benzylpenicillin or amoxicillin
Epiglottitis	H. influenzae (type b)	Cefotaxime
Dental abscess, gingivitis	Anaerobes, streptococci	Metronidazole or co-amoxiclav
Conjunctivitis	S. aureus, S. pneumoniae, H. influenzae	Chloramphenicol (topical)

Skin and subcutaneous tissue infections

Cellulitis		
Normal host	S. pyogenes, S. aureus	Benzylpenicillin and flucloxacillin, or clindamycin
Diabetic	As above, also anaerobes, enterococci, 'coliforms'	Co-amoxiclav; or cefotaxime, ampicillin and metronidazole
Leg ulcer	S. aureus, anaerobes, S. pyogenes, group B β-haemolytic Streptococcus	Local cleansing, flucloxacillin or clindamycin

Continued on p. 22.

Table 3.1 *(continued)*

Infection/age of patient	Causes	Recommended antibiotics
Wound infections		
Associated with:		
clean surgery	*S. aureus*	Flucloxacillin
dirty surgery	Anaerobes, *S. aureus*, 'coliforms'	Co-amoxiclav, or cefotaxime and metronidazole
Necrotizing fasciitis	*S. pyogenes*, other streptococci, *S. aureus*, anaerobes, 'coliforms'	Benzylpenicillin, gentamicin and metronidazole (± flucloxacillin)
Bite	*S. aureus*, *Pasteurella multocida*, anaerobes	Co-amoxiclav
Cannula-related	*S. aureus*	Flucloxacillin
Septicaemia (origin/association)		
Unknown	Gram-positive, Gram-negative and/or anaerobes	Ampicillin, gentamicin and metronidazole; or piperacillin/tazobactam; or meropenem
IV cannula	*S. aureus*, *S. epidermidis*	Vancomycin and flucloxacillin
Urinary tract	'Coliforms', *P. aeruginosa*	Cefuroxime or ciprofloxacin
Biliary tract	'Coliforms', anaerobes, enterococci	Cefuroxime and metronidazole, or piperacillin/tazobactam, or meropenem
Skin, subcutaneous tissue	*S. aureus*, *S. pyogenes*, anaerobes, *C. perfringens*	Benzylpenicillin, flucloxacillin, metronidazole, cefotaxime
Neutropenia	'Coliforms', *P. aeruginosa*	Gentamicin and piperacillin
Neonatal	'Coliforms', *Listeria*, group B β-haemolytic *Streptococcus*	Gentamicin and ampicillin

System-Based Infections

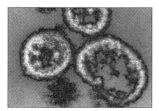

Eye and Upper Respiratory Tract Infections

Clinical syndromes

Conjunctivitis, keratitis and uveitis

Major presenting feature: *painful red eye.*

Causes

Conjunctivitis
- ADENOVIRUS
- CHLAMYDIA TRACHOMATIS
- *Streptococcus pneumoniae, Neisseria gonorrhoeae, Staphylococcus aureus, Haemophilus influenzae, N. meningitidis,* leptospirosis
- Enteroviruses, herpes simplex, varicella zoster, measles, rubella, *Mycobacterium tuberculosis,* Stevens–Johnson syndrome.

Keratitis
- HERPES SIMPLEX
- *Staphylococcus aureus, Streptococcus pneumoniae, Streptococcus pyogenes, Pseudomonas aeruginosa, N. gonorrhoeae*
- Varicella zoster, measles, adenovirus, mumps
- Fusarium, acanthamoeba, *Treponema pallidum, M. tuberculosis, M. leprae.*

Uveitis
- Herpes simplex, varicella zoster

- *T. pallidum, N. gonorrhoeae,* brucellosis, toxoplasmosis, *M. tuberculosis, M. leprae,* Lyme disease, typhoid, leptospirosis.

Distinguishing features
See Table 4.1.

Complications
Conjunctivitis: ophthalmia neonatorum, keratitis, uveitis.
Keratitis: corneal scarring, ulceration and perforation (leading to blindess), hypopyon, uveitis.
Uveitis: hypopyon, synechiae, raised intraocular pressure.

Investigations
- Full blood count (FBCt), differential white cell count (WCCt) and erythrocyte sedimentation rate (ESR)
- Chest X-ray (CXR) (for evidence of hilar lymphadenopathy indicating tuberculosis or sarcoidosis)
- Conjunctival/corneal scrapings for cytology
- Swabs for bacterial, viral, chlamydial and fungal culture
- Swab for adenovirus polymerase chain reaction (PCR)
- Acute and convalescent serology for infectious agents (as above).

	Keratitis	Conjunctivitis	Uveitis
Visual acuity	Impaired	Normal	Impaired
Pain	Severe	Gritty	Severe
Discharge	Only if bacterial	Present	Absent
Photophobia	Common	Uncommon	Common
Blepharospasm	Common	Uncommon	Rare
Ciliary infection	Present	Absent	Present
Cornea	Ulcer/oedema	Clear	Clear
Pupil	Normal	Normal	Normal/small, irregular
Pupillary reaction	Normal	Normal	Sluggish

Table 4.1 Distinguishing features of red eye.

Differential diagnoses
Acute glaucoma, subconjunctival haemorrhage, foreign body, trauma, episcleritis, scleritis.

Treatment
Conjunctivitis
Bacterial: topical antibiotics (e.g. chloramphenicol), usually sufficient for bacterial infection.
Chlamydial: topical tetracycline or erythromycin, and oral co-trimoxazole, erythromycin (or other macrolide) or tetracycline.

Keratitis
Bacterial: subconjunctival and parenteral drugs usually needed in addition to intensive topical antibiotics.
Herpetic: oral acyclovir.

Uveitis
Herpetic: oral acyclovir, topical corticosteroid and mydriatic.

Prevention
• Erythromycin and tetracycline are effective in preventing ophthalmia neonatorum due to *N. gonorrhoeae* and *C. trachomatis*: silver nitrate protects only against *N. gonorrhoeae*. These agents can be used where there is high endemicity for either infection.
• Children born to mothers with known gono-coccal infection should be treated with parenteral antibiotics (e.g. benzylpenicillin).

Tonsillitis/pharyngitis
Major presenting feature: *sore throat.*

Causes
• *STREPTOCOCCUS PYOGENES*
• EPSTEIN–BARR VIRUS (EBV)
• *CORYNEBACTERIUM DIPHTHERIAE*
• *Mycoplasma pneumoniae*, groups C and G β-haemolytic streptococci, *Corynebacterium haemolyticum*, *Corynebacterium ulcerans*, *C. trachomatis*, fusobacterium, *N. gonorrhoeae*, *N. meningitidis*, Toxoplasma
• Adenovirus, coxsackievirus, herpes simplex, parainfluenza, cytomegalovirus (CMV), human immunodeficiency virus (HIV).

Distinguishing features
See Table 4.2.

Complications
Streptococcus pyogenes: scarlet fever, quinsy, rheumatic fever, glomerulonephritis, otitis media, erythema nodosum, erythema multiforme, Henoch–Schönlein purpura.
Epstein–Barr virus: haemolytic anaemia, thrombocytopenia, airways obstruction (tonsillar), splenic rupture, hepatitis, Guillain–Barré syndrome, rash with antibiotics, meningoencephalitis.
Corynebacterium diphtheriae: myocarditis, neuropathy (palate (3 weeks), ocular (4

	S. pyogenes	Glandular fever	C. diphtheriae
Age group	Children	Teenagers	Children and young adults
Spread	Droplets	Saliva exchange	Droplets
Incubation period	2–5 days	14–42 days	2–5 days
Onset	Abrupt	Slow	Abrupt
Tonsillar exudate	White, thin	White, thin	Grey-green, thick, with possible spread to palate, pharynx or larynx
Palate	Normal	Petechiae	Membrane (as above)
Lymph nodes	Tender, no oedema	Non-tender, no oedema	Non-tender, oedema (bullneck)
Fever	High	High	Low grade
Toxicity	Moderate	Moderate	Marked
White cell count	Raised, neutrophilic	Raised, lymphocytic	Raised, neutrophilic
Heterophile antibody	Negative	Positive	Negative
Throat swab culture	Positive	Negative	Positive

Table 4.2 Distinguishing features of sore throat.

weeks), respiratory (7 weeks), peripheral (10 weeks)), airways obstruction (laryngeal), glomerulonephritis.

Investigations
• FBCt with differential WCCt
• Heterophile antibody test (Paul–Bunnell or Monospot)
• Throat swab for S. pyogenes: if negative, acute and convalescent sera for antistreptolysin O (ASO) titre
• If diphtheria possible: throat swab for C. diphtheriae (telephone laboratory); electrocardiogram (ECG) to detect myocarditis
• Throat swab for viral culture
• Blood culture (for Fusobacterium necrophorum)
• CXR for lung abscesses (F. necrophorum) if systemically unwell.

Differential diagnoses
If tonsillar or nearby swelling: quinsy, lymphoma, leukaemia, tuberculosis, cancer, retropharyngeal abscess.

Treatment
Streptococcus pyogenes
• Intravenous (IV) benzylpenicillin if severe, oral amoxicillin or penicillin V if mild to moderate
• Local analgesia (benzydamine or aspirin gargle).

Epstein–Barr virus
• Local analgesia (benzydamine or aspirin gargle)
• Consider IV hydrocortisone if upper airways obstruction or toxicity.

Corynebacterium diphtheriae
• Intramuscular (IM)/IV antitoxin
• IV erythromycin or benzylpenicillin
• Bed rest, cardiac monitoring.

Prevention
Streptococcus pyogenes
• Penicillin V for patients with a history of rheumatic fever.

Corynebacterium diphtheriae
• Immunization at 2, 3 and 4 months and at 4 years

- Treatment of carriers with oral erythromycin
- Postexposure erythromycin prophylaxis for close contacts.

Parotitis and cervical lymphadenitis

Major presenting feature: *swelling and pain around the angle of the jaw.*

Causes

Parotitis
- MUMPS
- STAPHYLOCOCCUS AUREUS
- Parainfluenza 3, coxsackievirus, CMV, *S. pyogenes*
- HIV.

Cervical lymphadenitis
- *Streptococcus pyogenes*
- *Staphylococcus aureus*
- EBV, toxoplasmosis, HIV, *Eikenella corrodens*, *C. diphtheriae*, anaerobes, cat-scratch disease, *Bacillus anthracis*, atypical mycobacteria, *M. tuberculosis*, CMV, Kawasaki disease.

Distinguishing features
See Table 4.3.

Table 4.3 Distinguishing features and causes of angle of jaw swelling.

Complications
Mumps: meningitis, epididymo-orchitis, encephalitis, pancreatitis, oophoritis.

Suppurative parotitis: septicaemia, massive neck oedema, respiratory obstruction, osteomyelitis.

Cervical lymphadenitis: suppuration, septicaemia, jugular venous and intracranial sinus thromboses.

Investigations
- FBCt with differential WCCt
- Amylase
- Throat swab for bacterial and viral culture
- Parotid duct swab for bacterial culture
- Blood culture
- CXR (to look for hilar lymphadenopathy)
- Serology: mumps S and V antigens; viral screen; ASO titre.

Differential diagnoses
Parotitis
Lymphoma, leukaemia, sarcoidosis, alcoholic liver disease, sialadenitis, salivary tumour, Sjögren syndrome, postoperative swelling, branchial cyst.

Cervical lymphadenitis
Lymphoma, leukaemia, cancer, sarcoidosis.

	Mumps	Suppurative parotitis	Cervical lymphadenitis
Incidence	Decreasing	Uncommon	Common
Age group	Children	All ages	Children
Spread	Droplet or saliva	Primary oral sepsis	Primary oral or skin sepsis
Incubation period	17–19 days	2–4 days	2–4 days
Onset	Subacute/acute	Acute	Acute
Parotid duct	Red	Exuding pus on massage	Normal
Angle of the jaw	Obscured	Obscured	Palpable
Tenderness	Moderate	Marked	Marked
Overlying skin	Normal	Red	Red/normal
White cell count	Normal/low, lymphocytic	Raised, neutrophilic	Raised, neutrophilic
Amylase	Raised	Normal	Normal

Treatment

Mumps
• Symptomatic: analgesics, mouth washes.

Suppurative parotitis
• IV benzylpenicillin and flucloxacillin
• Surgical drainage may be needed.

Cervical lymphadenitis
• IV benzylpenicillin and flucloxacillin
• Surgical drainage may be needed if suppuration develops.

Prevention

Mumps
Immunization with live mumps vaccine as part of measles, mumps and rubella (MMR) administration.

Epiglottitis and laryngotracheobronchitis
Major presenting feature: *difficulty in breathing*.

Causes

Epiglottitis
• HAEMOPHILUS INFLUENZAE (TYPE B)
• *Staphylococcus aureus*, non-B capsulate *H. influenzae*, *S. pneumoniae*, *S. pyogenes*, varicella zoster.

Table 4.5 Distinguishing features of croup and epiglottitis.

Laryngotracheobronchitis (croup)
• PARAINFLUENZA I
• Parainfluenza 2 or 3, measles, respiratory syncytial virus (RSV), influenza, adenovirus, rhinovirus, *C. diphtheriae*, *M. pneumoniae*.

Distinguishing features
See Table 4.5.

Complications
Epiglottitis: airways obstruction, hypoxia, septicaemic shock, vagally induced cardiopulmonary arrest.
Laryngotracheobronchitis: respiratory failure, pulmonary oedema, pneumothorax, aspiration pneumonia.

Investigations
An anaesthetist must be called immediately if epiglottitis is suspected. No investigations should be performed until the personnel and equipment to secure the airway are available.
• FBCt with differential WCCt
• Blood culture (septicaemia common with epiglottitis)
• Throat swab for bacterial and viral culture
• Epiglottal culture (*only* at intubation)
• Nasopharyngeal aspirate for rapid viral antigen detection (RSV, parainfluenza 1–3, adenovirus)

	Viral croup	Epiglottitis
Incidence	Common	Rare
Season	Winter	All year
Age group	< 3 years	2–5 years
History	Days	Hours
General condition	Agitated	Lethargic/toxic
Fever	Low grade	High
Toxaemia	Absent	Present
Cough	Brassy	Weak/absent
Audible stridor	Marked	Moderate
Drooling saliva	Absent	Present
White cell count	Normal/low, normal/lymphocytic	High, neutrophilic
X-ray of neck	Subglottic swelling	Epiglottic swelling

- If diphtheria possible (immigrant, throat features): throat swab for *C. diphtheriae* (telephone laboratory); ECG
- Capillary/arterial gases or oxygen saturation
- Chest X-ray
- Lateral neck X-ray.

Differential diagnoses

Foreign body, allergic reaction, retropharyngeal abscess, suppurative tracheitis, laryngomalacia, recurrent laryngeal nerve paralysis.

Treatment

Epiglottitis

- Secure airway (endotracheal tube or tracheostomy)
- Intensive care monitoring
- Cefotaxime (because *H. influenzae* type b resistance to ampicillin is 30% and to chloramphenicol is 1–3% neither can be used confidently).

Laryngotracheobronchitis

- Supplemental oxygen
- Humidification
- Corticosteroids (nebulized or oral) should be used in moderate to severe cases.

Prevention

Epiglottitis

- Rifampicin prophylaxis for all family contacts where there is a child < 5 years
- *Haemophilus influenzae* type b (Hib) vaccine at 2, 3 and 4 months.

Bacterial sinusitis

Major presenting features: *facial pain, blocked nose, fever, postnasal drip.*

Causes

Predisposing

- Viral infection, cystic fibrosis, Kartagener's syndrome (immotile cilia), nasal polyps, allergic rhinitis, foreign body, septal deviation, immune deficiency, dental abscess, nasopharyngeal tumours, nasal packing.

Bacterial — acute

- HAEMOPHILUS INFLUENZAE, S. PNEUMONIAE, S. pyogenes, *Moraxella catarrhalis*, *S. aureus*.

Bacterial — chronic

- ANAEROBES, STREPTOCOCCUS MILLERI, 'coliforms', *P. aeruginosa*
- *H. influenzae*, *S. pneumoniae*, *S. pyogenes*, *M. catarrhalis*, *S. aureus*
- *Mycobacterium tuberculosis*, fungi.

Complications

Chronic sinusitis, meningitis, intracranial abscess, subdural empyema, osteomyelitis.

Investigations

- Sinus X-ray (for mucosal thickening or fluid level)
- Imaging with computerized tomography (CT) or magnetic resonance imaging (MRI) scan (usually reserved for complicated disease)
- Antral lavage (for chronic disease): specimens for microscopy, Gram and Ziehl–Neelsen stain; bacterial, mycobacterial and fungal culture
- Antroscopy and biopsy (for chronic disease).

Differential diagnoses

Chronic hyperplastic allergic sinusitis, sarcoidosis, malignancy, upper molar dental abscess.

Treatment

- Antibiotics:
  - acute sinusitis: amoxicillin, erythromycin (or a newer macrolide) — or co-amoxiclav
  - chronic sinusitis: co-amoxiclav
- Decongestants
- Surgery (for chronic disease): antral lavage or antrostomy (simple and radical)
- Topical corticosteroids (if allergic component).

Prevention

- Surgery (for polyps, deviated nasal septum).

Bacterial otitis media

Major presenting features: *earache, fever, ear discharge, deafness.*

Causes
Predisposing
• Viral infection, enlarged adenoids, immune deficiency, nasopharyngeal tumours.

Bacterial — acute
• Haemophilus influenzae, S. pneumoniae, S. pyogenes, M. catarrhalis, S. aureus
• Mycoplasma pneumoniae, C. trachomatis.

Bacterial — chronic
• Anaerobes, 'coliforms', S. milleri, P. aeruginosa
• Haemophilus influenzae, S. pneumoniae, S. pyogenes, M. catarrhalis, S. aureus
• Mycobacterium tuberculosis, fungi.

Complications
'Glue ear' (persistent middle ear effusion), mastoiditis, labyrinthitis, meningitis, intracranial abscess, facial nerve paralysis, chronic perforation of the drum, cholesteatoma.

Investigations
• Ear swab for bacterial and fungal culture if chronic perforation
• Throat swab for S. pyogenes.

Differential diagnoses
Viral infection, otitis externa, foreign body, 'glue ear'.

Treatment
• Paracetamol suspension
• Antibiotics:
  • acute otitis media: amoxicillin, erythromycin (or a newer macrolide), or co-amoxiclav
  • chronic otitis media: co-amoxiclav
• For complicating 'glue ear':
  • myringotomy
  • grommet insertion.

Prevention
• Adenoidectomy (indicated for children with prolonged glue ear).

Specific infections

Trachoma and inclusion conjunctivitis
Epidemiology
Trachoma
Trachoma is caused by *Chlamydia trachomatis*, and is the commonest infective cause of blindness in the world. It is endemic in the hot, dry areas of the tropics and subtropics.
• Transmission is from an infected eye by hands, materials and flies
• Young children are most at risk
• It is associated with inadequate public and personal hygiene
• Repeated infections occur: PCR suggests many are infected chronically
• Incidence is now falling but it has been estimated it has been the cause of > 20 million cases of blindness globally
• It remains a common cause of preventable blindness in Africa, the Middle East and parts of Asia
• The incubation period (IP) is 5–7 days.

Inclusion keratoconjunctivitis
Inclusion keratoconjunctivitis is common in temperate climates and is also caused by *C. trachomatis*.
• Transmission is usually sexual (adults), vertical (ophthalmia neonatorum) or by direct or indirect contact (children)
• There is often associated chlamydial genital infection in the patient or partner
• The IP is 5–12 days.

Pathology and pathogenesis
Chlamydia are intracellular pathogens dependent on the host cell for energy. They have two major forms, the infectious elementary body and the non-infective reticulate body. *Chlamydia trachomatis* is divided into 15 serovars; of these:
• A, B and C cause trachoma
• D, E, F and G cause genital infections, ophthalmia neonatorum and adult ocular infections in developed nations

• LI, L2 and L3 cause lymphogranuloma venereum.

In trachoma an acute inflammatory response with purulent conjunctivitis and follicular reaction in the superior tarsal conjunctiva follows infection. Fibrous tissue and new blood vessels (pannus) form with repeated infections, leading to blindness. The eyelids become thickened and everted, leaving the conjunctivae susceptible to damage from infection and dust. In inclusion conjunctivitis, follicles are more pronounced on the lower tarsal conjunctiva and scarring is rare.

Clinical features
• Acute purulent follicular conjunctivitis
• Lachrymal gland and preauricular gland enlargement.

Complications
Trachoma
• Recurrent bacterial secondary infections
• Corneal scarring, new vessel formation
• Lid eversion
• Blindness.

Inclusion conjunctivitis
• Corneal inflammation with infiltrates, erosions and some new vessel formation.

Diagnosis
• Antigen or inclusion bodies on conjunctival smears
• Culture
• PCR for chlamydial DNA
• Serology is unhelpful, but measurement of tear antibody may be useful.

Treatment
• For the acute attack, treatment with tetracycline eye ointment and/or oral tetracycline or azithromycin is effective and helps to prevent secondary cases. Topical therapy alone may not eradicate infection
• Surgery to correct lid deformities may prevent blindness in trachoma
• Treatment of sexual partners in adult inclusion conjunctivitis.

Prevention
• Mass treatment with tetracycline ointment or oral azithromycin
• Improving personal hygiene and general sanitation.

Streptococcal tonsillopharyngitis
Epidemiology
Sore throats are common in the general population and S. *pyogenes* is the cause in one-quarter of cases.
• Infection is usually sporadic, but periodic outbreaks occur
• The peak incidence is in school-age children during winter
• Transmission is by upper respiratory droplets and is facilitated by overcrowding. Rarely, milk-borne outbreaks occur
• *Streptococcus pyogenes* pharyngeal colonization occurs in 15–20% of children; in adults it is considerably lower
• Groups C and G β-haemolytic streptococci occasionally cause tonsillitis
• The IP is 2–5 days.

Pathology and pathogenesis
Streptococcus pyogenes contains or produces a variety of intra- and extracellular toxins which enhance virulence. The secreted proteins include:
• pyrogenic exotoxins A, B and C (responsible for scarlet fever and toxaemia)
• streptolysins O and S (responsible for β-haemolysis)
• deoxyribonucleases, hyaluronidase and streptokinase which facilitate the spread of bacteria through tissues.
Antibodies to these products are used in the serodiagnosis of recent streptococcal infection. Of the intracellular and cell wall products, the M protein is the major virulence antigen. It is also the major antigen by which strains are typed, and over 80 such serotypes are recognized. This is epidemiologically important because certain serotypes are associated with streptococcal diseases:
• scarlet fever (types 1, 3 and 4)

- glomerulonephritis (types 12, 49, 55, 57 and 60)
- rheumatic fever (types 1, 3, 6 and 18).

Clinical features

Distinction of streptococcal tonsillitis from other causes is often difficult. Suppurative and postinfectious complications of streptococcal infection can be prevented or attenuated by antibiotics, whereas their inappropriate use in glandular fever may result in a severe skin rash. It is therefore important to make a clinical diagnosis. Typical of streptococcal tonsillopharyngitis are:

- an abrupt onset with sore throat, fever and headache
- vomiting and abdominal pain in children
- white exudate over swollen tonsils
- tender upper anterior cervical lymphadenitis
- a self-limiting illness in treated, uncomplicated cases lasting 3–5 days
- the occurrence of a scarlitiniform rash in 5–10% of cases.

Complications

Suppurative
- Peritonsillar (quinsy) and retropharyngeal abscesses
- Otitis media and sinusitis
- Suppurative cervical lymphadenitis
- Jugular venous and intracranial sinus thromboses
- Pneumonia.

Non-suppurative
- Scarlet fever
- Streptococcal toxic shock
- Rheumatic fever
- Glomerulonephritis
- Erythema nodosum
- Henoch–Schönlein purpura.

Diagnosis

Most sore throats are viral. Exudative tonsillitis may also be due to *C. diphtheriae* (now rare), EBV and adenovirus (when it may be associated with conjunctivitis—pharyngoconjunctival fever). Diagnosis is confirmed by:

- throat swab culture on blood agar (β-haemolysis and reaction with group A-specific antibody)
- a fourfold rise or fall or single high-titre antibody to one of the major antigenic constituents (usually ASO antibodies).

Rapid antigen detection using kits is highly specific but relatively insensitive (75%) compared to throat culture.

Treatment

Therapy is intended to prevent complications. It has little effect on the course of the tonsillitis. First-line treatment is penicillin: group A streptococci remain universally susceptible. Where there is a history of penicillin allergy, erythromycin is a suitable alternative, although 5% of strains are resistant. When the clinical features are severe or suppurative complications have intervened, IV benzylpenicillin should be administered.

Prevention

Recurrent tonsillitis
- Tonsillectomy is not of benefit except in a few children with recurrent, bacteriologically proven infections.

Rheumatic fever
- Initial treatment course of 10 days of oral penicillin
- Continuous prophylaxis with oral penicillin.

Prognosis

Fatalities are exceptionally rare but may occur in toxin-mediated (scarlet fever and toxic shock syndrome) and severe suppurative (septicaemia and intracranial sinus thromboses) complications.

Diphtheria

Epidemiology

Diphtheria is a severe infection characterized by membrane formation in the throat and toxaemia which damages heart muscle and nerve tissue. *Corynebacterium diphtheriae* is characterized as follows.

- It is differentiated into three bacteriological

types — *gravis*, *intermedius* and *mitis*; all are capable of severe disease
• It is transmitted by droplet from a naso-pharyngeal case or carrier; occasionally skin infection occurs when direct contact is important
• It is now rare in the UK (>90% immunized, with <5 cases/year). In unimmunized popula-tions it can cause devastating epidemics
• It is mainly a disease of under-resourced nations, but an epidemic in the former Soviet Union occurred in the late 1990s
• It rarely affects adults unless unimmunized; most cases are imported
• It can cause infection of skin (cutaneous diph-theria) and rarely be invasive (injection drug users)
• It has an IP of 2–5 days.

Pathology and pathogenesis

Corynebacterium diphtheriae is relatively non-invasive and causes a mild inflammatory reaction in the tonsil. Virulence results from the production of a potent exotoxin which inhibits protein synthesis by interference with mRNA. Locally this causes epithelial necrosis, an adherent membrane and surround-ing oedema. As more toxin is produced it is absorbed from the membrane into the bloodstream where it affects the heart and nerves. Toxin is readily absorbed from the throat but only slightly from the nose, larynx or skin. Non-toxigenic strains of *C. diphtheriae* ac-quire the ability to produce toxin if infected with bacteriophages encoding the toxin gene (phage conversion).

Clinical features

Clinical features depend upon the site of mem-brane formation. In tonsillopharyngeal disease (>50% of cases):
• sore throat and low-grade fever commence gradually
• the membrane first appears on one or both tonsils and may spread to the pharynx, palate or buccal mucosa
• the membrane is off-white and thick; the

extent of membrane formation correlates with severity
• there may be associated marked cervical adenitis and oedema, producing the classical 'bullneck'.
In laryngeal disease:
• hoarseness, a croupy cough and stridor develop
• with time, inspiratory recession of tissues and cyanosis will occur.
In anterior nasal disease:
• blood-stained unilateral nasal discharge occurs
• symptoms of toxicity are mild.
In cutaneous infection:
• chronic cutaneous ulcers develop, with a grey membrane
• symptoms of toxicity are mild
• the ulcers are a reservoir of *C. diphtheriae* that can lead to pharyngeal infection/carriage.

Complications

• Respiratory obstruction from pseudomem-branes
• Myocarditis
• Neuritis
• Pneumonia
• Glomerulonephritis.
Two-thirds of patients develop myocarditis but in only 10–25% of these is this clinically impor-tant. It occurs between days 10 and 20; the ECG may reveal ST–T wave changes and heart block. Neuritis affects 75% of patients with severe dis-ease and results from demyelination. It afflicts cranial and peripheral nerves in a sequential order: palate (3 weeks), oculomotor (4 weeks), respiratory (7 weeks) and peripheral (10 weeks). All paralyses recover in time.

Diagnosis

The major differential diagnoses are glandular fever and streptococcal tonsillitis. Laryngeal diphtheria can be confused with epiglottitis, croup and foreign body obstruction. Confirma-tion of diagnosis is by culture from throat and nasal swabs but treatment must not be delayed. Because non-toxigenic *C. diphtheriae* are not infrequently recovered from throat and nasal swabs, toxin studies must be performed on the isolate.

Treatment

The outcome is improved by prompt initiation of antitoxin therapy using horse hyperimmune antiserum. The dose depends on the site and severity of disease. Anaphylactic reactions may occur, especially in patients previously treated with this product. Penicillin or erythromycin should be given to eradicate *C. diphtheriae* and post-treatment clearance swabs checked. Sensitivies must be checked because occasional resistance to erythromycin occurs.

Cardiac monitoring is essential for anything more than mild disease and patients should be rested in bed.

Prevention

- Immunization at 2, 3 and 4 months, and at entry into primary school
- Close contacts of the index case should have:
  - their throats examined by an experienced physician
  - nose and throat swabs taken for *C. diphtheriae* culture
  - a course of erythromycin prophylaxis (a newer macrolide, rifampicin or clindamycin are alternatives)
  - their immunization status assessed. If non-immune or of uncertain immunity, they should receive a primary course of diphtheria vaccine. If previously immunized, they should receive a reinforcing dose of adult diphtheria vaccine
- The Schick test is no longer available for determining immunity
- The patient should be nursed in isolation.

Prognosis

Case fatality rate is 5–10%. The causes of death are severe toxaemia or laryngeal obstruction in the early days, heart failure in the 2nd or 3rd week, or respiratory failure at the 6th week.

Glandular fever

Epidemiology

Infection with EBV is almost universal. In childhood, primary EBV infection is usually non-specific or asymptomatic, whereas in adolescents and adults glandular fever usually ensues. In EBV infection:

- 50% of children in Western countries are infected by 5 years of age
- earlier acquisition occurs in the poor and developing nations
- the virus is excreted intermittently in most seropositive persons following primary infection
- transmission is by direct or indirect (via hands) saliva contact; rarely, infection follows blood transfusion.

In glandular fever:

- sporadic disease occurs with equal sex distribution
- the incidence is 50/100 000 in Western countries
- only 6% have a history of contact with another case
- the IP is 2–6 weeks.

Pathology and pathogenesis

EBV is a herpesvirus. After oropharyngeal inoculation, EBV infects local epithelial cells and B lymphocytes. Dissemination of infected B lymphocytes then follows, with subsequent production of a T-cell response characterized by atypical mononucleosis. This results in lymphoid hyperplasia with lymphadenopathy, splenomegaly and hepatomegaly. After infection, cells that contain the EBV genome are capable of continuous *in vitro* cultivation (immortalization).

Clinical features

The classic triad of sore throat, fever and lymphadenopathy follows on from a prodrome of fever, tiredness, vague malaise and headache. The important clinical findings are:

- exudative tonsillitis with peritonsillar oedema
- petechiae at the junction of hard and soft palate

• cervical, axillary and inguinal lymphadenopathy with discrete non-tender glands
• splenomegaly (50%), hepatomegaly (15%) and jaundice (5%)
• maculopapular pruritic rash (90%) after antibiotics (especially ampicillin/amoxicillin). Occasionally a rash may occur without preceding antibiotics (5%).

Complications

The vast majority of patients with glandular fever recover uneventfully. Complications are rare but include:
• pharyngeal and tracheal obstruction
• Coombs' positive haemolytic anaemia, thrombocytopenia
• Guillain–Barré syndrome, meningoencephalitis
• splenic rupture, severe hepatitis, myocarditis and pneumonitis
• chronic fatigue.
True relapses of glandular fever are rare.

Diagnosis

Characteristic haematological findings are leucocytosis, relative and absolute lymphocytosis, the presence of atypical lymphocytes, and a positive heterophile antibody test—Paul–Bunnell or Monospot (90% of cases). When considering the diagnosis:
• the heterophile antibody test may be negative early in infection, in children, in the elderly and in mild cases
• differential absorption with guinea-pig kidney cells and ox red cells distinguishes glandular fever from heterophile antibodies associated with serum sickness
• false-positive heterophile antibody tests occasionally occur in hepatitis A, malaria, rubella and lymphoma
• positive heterophile antibody tests persist for many months
• atypical lymphocytosis also occurs in CMV, HIV, human herpesvirus 6 (HHV-6), mumps, toxoplasma and hepatitis virus infections
• heterophile antibody-negative glandular fever may result from EBV, HIV, CMV or *Toxoplasma* infections

• positive IgM serology to viral capsid antigen confirms the diagnosis. This is particularly useful in confirming the diagnosis in heterophile antibody-negative cases
• mildly deranged transaminases occur in 90%.

Treatment

A short course of steroids reduces tonsillopharyngeal oedema and is indicated when there is impending airways obstruction. They may also be useful when severe thrombocytopenia or acute haemolysis complicates infection.

Prevention

• Isolation in hospital is unnecessary
• No vaccine is available.

Prognosis

Practically all cases make a full recovery. Death in an immunocompetent person is rare but may result from respiratory obstruction, ruptured spleen or encephalitis. Fatalities may also occur in patients with inherited T-cell defects. EBV has been associated with several malignancies and other HIV-related conditions:
• nasopharyngeal carcinoma in China
• Burkitt's lymphoma in equatorial Africa
• immunoblastic lymphoma in X-linked lymphoproliferative syndrome
• B-cell lymphomas in the immunocompromised
• primary central nervous system (CNS) lymphoma, non-Hodgkin's lymphoma, Hodgkin's disease, oral hairy leucoplakia and lymphocytic pneumonitis in HIV-infected patients
• Castleman's disease.

Mumps
Epidemiology

Mumps is a generally mild and self-limiting acute infection caused by a member of the paramyxovirus group. Humans are the only natural hosts and infection is common and widespread.
• 85% of adults show serological evidence of past infection
• Schoolchildren and adolescents are principally affected, although more cases are now being seen in adults

• Clinical mumps is rare in children < 2 years of age
• Cyclical (2–5 years) and seasonal (spring) peaks occur
• Transmission is by droplet spread or direct/indirect contact with saliva
• Over 95% reduction in incidence has occurred where attenuated mumps vaccine has been in use for 10 years
• Infectivity lasts from a week before salivary gland swelling starts to up to 9 days after, with a peak just before and at the onset of the parotitis
• The IP is 17–19 days, with limits of 12–25 days
• Clinical disease gives lifelong immunity.

Pathology and pathogenesis

Initially the virus replicates in the epithelium of the upper respiratory tract, followed by viraemic dissemination to other organs including salivary glands and meninges. Oedema and mononuclear infiltration are seen.

Clinical features

The illness presents with a non-specific prodrome of headache, sore throat, malaise and fever lasting 1–2 days. Salivary gland inflammation and enlargement then follows. Parotitis:
• is present in 70% of cases
• starts unilaterally but involves the other gland after 1–5 days in 75% of cases
• is heralded by earache and discomfort/tenderness at the angle of the jaw, which is aggravated by chewing
• causes rapid, painful and tender enlargement, obscuring the angle of the jaw
• is associated with involvement of the other salivary glands in 10%
• resolves over 1–2 weeks
• may result in difficulty in opening the mouth (trismus)
• is associated with swelling and redness at the opening of the parotid duct opposite the upper second molar.

Complications

Neurological
• Meningitis
• Encephalitis
• Nerve deafness, facial palsy, Guillain–Barré syndrome, transverse myelitis.

Mumps meningitis is an important cause of viral meningitis and may be the presenting syndrome, when only half have clinical evidence of parotitis. It occurs in 10% of those presenting with parotitis and can precede salivary gland involvement, or follow it by as long as 2 weeks; usually it starts 4 days after parotitis. It is associated with a lymphocytic cerebrospinal fluid (CSF) with raised protein, normal or low (5–10%) glucose and mumps virus isolation. Encephalitis is rarer, usually presents in the later stage of infection, and has a significant morbidity and mortality.

Glandular
• Epididymo-orchitis, oophoritis
• Pancreatitis.

Epididymo-orchitis occurs in 25% of postpubertal males and is unilateral in 85% and appears as other signs settle. It is heralded by the abrupt return of fever and malaise and indicated by rapid painful swelling of the testes and reddening of the overlying scrotal skin. Occasionally it is associated with atrophy of the testis; sterility is very rare. Rarely, it is the only manifestation of mumps. Oophoritis occurs much less frequently.

Other complications
• Arthritis, myocarditis, pericarditis.

Diagnosis

In typical cases, the diagnosis is clinical and simple, but in atypical cases or where there are associated complications, laboratory confirmation is necessary. Circumstantial evidence of recent mumps infection is supported by:
• a normal or low WCCt with a relative lymphocytosis
• an elevated serum amylase
• a lymphocytic CSF with raised protein and normal or low sugar.

Proof of infection is provided by:
• isolation of the virus from CSF, saliva or urine
• demonstration of a fourfold rise or fall, or single high-titre antibody to mumps 'S' (0–12 weeks) or 'V' (2 weeks–years) antigens.

Treatment
• Symptomatic — mild   analgesics,   mouth washes and easily chewed diet
• In orchitis, strong analgesics, local ice packs and scrotal support are comforting.

Prevention
• In hospital, patients should be isolated
• Immunization of susceptible contacts following exposure is not protective. Since subclinical infection frequently occurs, a non-immune contact should be considered infectious from 12 to 25 days after exposure
• Since 1988, attenuated live mumps vaccine as a component of combined MMR vaccine has been in routine use in the UK. The vaccine is recommended for all children aged 12–15 months unless contraindicated, and should also be offered to older, unimmunized children before they start school. The immunity appears to be long-lasting and the vaccine is well tolerated. Transient swelling of the parotid glands lasting for less than 24 h may develop in the 3rd week.

Prognosis
Apart from a very rare death from encephalomyelitis, there is no mortality from mumps. Permanent sterility or deafness are very rare sequelae.

Epiglottitis
Epidemiology
Acute epiglottitis is a rapidly progressive life-threatening infection resulting from capsulate *H. influenzae* type b. It is a medical emergency which:
• affects children of 2–5 years of age; occasionally, adults are affected
• is transmitted by aerosol from a carrier or a case

• has an IP of 1–4 days
• has become rare with the introduction of *H. influenzae* type b (Hib) vaccine.

Pathology and pathogenesis
Capsulate strains of *H. influenzae* are invasive by virtue of surface pili (aid adherence and mucosal penetration), lipopolysaccharide (inhibits ciliary beating) and the capsule (assists colonization and resistance to neutrophils). Infection is infrequent but invariably results in severe disease (meningitis, epiglottitis, septicaemia, cellulitis, pneumonia). By contrast, non-capsulate strains commonly cause local complications (otitis media, sinusitis, conjunctivitis, bronchopneumonia) but are very rarely invasive. Following colonization and local invasion, a brisk inflammatory reaction develops in the epiglottis, with rapid swelling resulting in airways obstruction.

Clinical features
Epiglottitis progresses rapidly and may be fulminant. Typically, children:
• have no prodromal illness
• present sitting, leaning forward and drooling secretions
• are unable to swallow or talk
• have a weak cough and inspiratory stridor
• are tachycardic, hypotensive, febrile and toxic-looking
• rapidly progress to airways obstruction.
The throat must *not* be examined until an anaesthetist is present and facilities are available to secure the airway. Attempts to do so may result in complete airway obstruction or vagally mediated cardiopulmonary arrest.

Complications
These may be local (airways obstruction, hypoxaemia) or systemic (septicaemic shock). Because complications are the rule and not the exception, the child must be admitted directly to the intensive care unit.

Diagnosis
The diagnosis is clinical. Typically, there is:

- a cherry-red, swollen epiglottis visible at intubation
- leucocytosis with neutrophilia
- *Haemophilus influenzae* type b grown from blood and epiglottal cultures
- a swollen epiglottis on lateral neck X-ray.

Treatment
- Establish the airway
- Give IV cefotaxime. Nearly one-third of *H. influenzae* type b isolates are resistant to ampicillin and 1–3% to chloramphenicol.

Prevention
- Hib vaccine at 2, 3 and 4 months
- The index case should receive rifampicin 20 mg/kg/day daily for 4 days before hospital discharge
- Close family contacts where there is a child <5 years should also receive rifampicin.

Prognosis
With early recognition, full recovery is usual. However, 5–10% of cases are fulminant and are often dead on arrival at hospital.

Laryngotracheobronchitis
Epidemiology
Acute laryngotracheobronchitis (croup) is an acute infection of the respiratory tract, particularly involving the subglottic area, resulting in the characteristic stridor.
- Most cases result from human parainfluenza viruses (HPIVs). Rhinoviruses, influenza, RSV and enteroviruses are occasionally responsible
- Four HPIVs have been identified; HPIV-1 and, less frequently, HPIV-2 are the leading causes of laryngotracheobronchitis (croup). HPIV-3 more frequently causes bronchiolitis
- HPIV is a single-stranded enveloped RNA virus. It is unstable in the environment, lasting only a few hours and is readily inactivated by disinfectants
- Disease is most commonly seen in children aged 6 months to 4 years. It is the cause of 10% of childhood respiratory illness

- 75% of children have serological evidence of HPIV-1 or -2 infection by 5 years
- HPIV occurs in winter epidemics in 2-yearly cycles, affects children between 3 months and 3 years of age, can cause second infections, and affects boys more commonly
- The infection is self-limiting; complications are rare
- The IP is 2–4 days.

Pathology and pathogenesis
The cardinal features of croup (stridor, hoarseness and cough) result from viral-induced inflammation and oedema of the larynx and trachea, although the whole respiratory tract is involved in the infection. Hypoxaemia results from parenchymal lung inflammation and laryngeal obstruction. Cilia damage predisposes to secondary bacterial infection.

Clinical features
Characteristics are as follows.
- There is a short prodrome of coryza, sore throat and cough
- Hoarseness, barking cough and stridor follow
- Airways obstruction develops with recession of the soft tissues of the neck and abdomen on inspiration and, in severe cases, cyanosis
- Symptoms are often worse at night
- Fever is not striking and the WCCt is normal or low
- After a fluctuating course lasting 4–5 days, the symptoms settle
- Airways obstruction is very rare in older children and adults.

Complications
- Respiratory failure, pneumothorax, pulmonary oedema and aspiration pneumonia
- Pulmonary collapse from thick secretions
- Bacterial superinfection
- Bronchiolitis obliterans
- Meningitis, encephalitis
- Airway hyperreactivity.

Differential diagnosis
- Epiglottitis, bacterial tracheitis, foreign body inhalation
- Bronchiolitis, bacterial pneumonia.

Diagnosis
The diagnosis is clinical. Viral aetiology is confirmed by:
- virus isolation from a nose or throat swab
- antigen or RNA detection from nasopharyngeal aspirates
- serology showing an IgM antibody response or a fourfold rise or fall in antibody titre.

Treatment
- Nurse in a humidified atmosphere
- Supplemental oxygen with monitoring of Po_2 or oxygen saturation
- Nebulized budesonide or oral prednisolone is beneficial for moderate to severe disease
- Endotracheal intubation or tracheostomy may be necessary to relieve laryngeal obstruction and remove bronchial secretions
- IV antibiotics for potential bacterial superinfection
- Aerosolized ribavirin has been used with possible benefit.

Prevention
- No vaccine is currently available
- Passively acquired maternal antibody may play a protective role
- Control of cross-infection in children's wards.

Prognosis
The illness is benign. Recovery is normally seen within 1–2 days. Fatalities result from respiratory failure or one of the other major acute complications.

Common cold
Epidemiology
The common cold is a mild self-limiting catarrhal illness associated with a low-grade fever. It is not a single entity but a clinical syndrome with many viral causes. It is a major cause of lost school and work days.
- Rhinoviruses account for one-third of cases; coronaviruses, adenoviruses, parainfluenza and influenza viruses, RSV and enteroviruses for another third; in the remaining it is impossible to identify the cause
- Secondary bacterial infection resulting in otitis media, sinusitis or tracheobronchitis occurs in 2–3%
- Reinfections are common and, on average, an adult suffers two or three, and a child six to eight colds per year
- Transmission is by droplet infection or direct contact, usually via hands
- Annual epidemics occur during the cold or wet seasons
- The IP is 12 h–3 days.

Pathology and pathogenesis
Rhinoviruses attach via specific receptors. Limited damage occurs following viral invasion of the columnar epithelial cells with rhinoviruses. CT scans have demonstrated thickened nasal walls, engorged turbinates and fluid or mucosal thickening in the sinuses, indicating that the common cold is not a localized infection. Pharyngitis and conjunctivitis may occur with adenoviral or enteroviral infection.

Clinical features and complications
The illness starts with slight fever, malaise and irritation of the nasal mucosa and pharynx, soon followed by a profuse watery nasal discharge, repeated sneezing and coughing. Later the nasal discharge becomes more purulent and the mild systemic symptoms subside. The only complication is secondary bacterial infection (H. influenzae, S. pneumoniae, S. pyogenes) leading to sinusitis, otitis media or tracheobronchitis.

Diagnosis, treatment and prevention
Rarely is there any indication to confirm the aetiology of a common cold, which is a clinical diagnosis. Nose and throat swabs for viral culture, direct antigen testing on nasopharyn-

geal aspirates and serology are available. Treatment is symptomatic with aspirin gargles for a sore throat, decongestants for a blocked nose and antibiotics only if secondary bacterial infection has occurred. Interferon A has a modest benefit on symptoms; there is no vaccine. Thorough hand-washing helps to interrupt transmission.

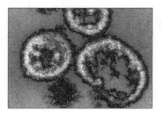

Lower Respiratory Tract Infections

Clinical syndromes

Pleural effusion, empyema and lung abscess

Major presenting features: *chest pain, cough, fever.*

Causes

Pleural effusion and empyema

Post-pneumonic: STAPHYLOCOCCUS AUREUS, STREPTOCOCCUS PNEUMONIAE, *Streptococcus pyogenes, Mycoplasma pneumoniae, Legionella pneumophila.*

Primary: MYCOBACTERIUM TUBERCULOSIS, STREPTOCOCCUS MILLERI, ANAEROBES, microaerophilic streptococci, *Escherichia coli* and other 'coliforms' (often polymicrobial), *Actinomyces, Nocardia.*

Lung abscess

Necrotizing pneumonia: S. AUREUS, *Klebsiella pneumoniae, Fusobacterium necrophorum, S. pyogenes, Pseudomonas aeruginosa.*

Aspiration: ANAEROBES, microaerophilic streptococci, *S. milleri* (often polymicrobial).

Others: M. TUBERCULOSIS, *Entamoeba histolytica,* actinomycosis, *Nocardia, Pseudomonas pseudomallei,* aspergilloma.

Distinguishing features
See Table 5.1.

Complications
Lung abscess: bronchopleural fistula, empyema, brain abscess, bronchiectasis.

Empyema: empyema necessitans, septicaemia, pericarditis.

Investigations
• Full blood count (FBCt), differential white cell count (WCCt), erythrocyte sedimentation rate (ESR)
• Chest X-ray (CXR) (Fig. 5.1): further imaging may be necessary, e.g. ultrasound scan/CT scan
• Blood culture (occasionally positive in empyema/lung abscess)
• Fluid from pleural cavity/lung abscess:
  • microscopy
  • Gram and Ziehl–Neelsen (ZN) stains
  • culture: aerobic and anaerobic; *M. tuberculosis*
  • protein concentration
  • antigen testing for *S. pneumoniae* (if negative culture).

	Pleural effusion	Empyema	Lung abscess
Fever	Normal/low	High/hectic	High/hectic
Chest pain	Present	Present	Absent
Clubbing	Absent	If chronic	If chronic
Chest X-ray			
Site	Costophrenic	Costophrenic or loculated	Lungs
Air/fluid level	Absent	Uncommon	Usually present
Full blood count			
White cell count	Normal	Raised, neutrophilic	Raised, neutrophilic
Anaemia	Absent	May be present	May be present
Pleural/abscess fluid			
White cell count increase	Moderate, lymphocytic	High, neutrophilic	High, neutrophilic
Protein	Moderate rise	Very high	Very high
Gram stain	Rarely positive	Usually positive	Usually positive
Culture	Negative	Positive	Positive

Table 5.1 Distinguishing features of intrathoracic fluid collections.

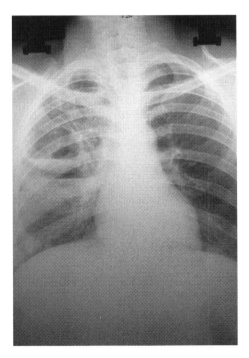

Fig. 5.1 Chest X-ray of a patient with anaerobic lung abscess showing a well-defined cavity with fluid level in the right upper lobe.

Differential diagnoses

Pleural effusion/empyema

• Neoplasm, pulmonary infarct, congestive cardiac failure, autoimmune disease (systemic lupus erythematosus (SLE) rheumatoid disease, etc.), hypoalbuminaemia, subdiaphragmatic infection, oesophageal perforation, pancreatitis.

Lung abscess

• Neoplasm, pulmonary infarct, infection distal to bronchial obstruction, septic embolus, Wegener's granulomatosis.

Treatment

Empyema and lung abscess

• Aspirate/drain fluid where feasible
• Initial parenteral antibiotics, guided by isolate(s) and sensitivity patterns
• Post-pneumonic: treat underlying condition:
  • *S. aureus*: flucloxacillin
  • *K. pneumoniae*: cefotaxime
  • *S. pyogenes*: benzylpenicillin
  • *M. pneumoniae*: erythromycin or a newer macrolide
  • *S. pneumoniae*: cefotaxime

- Primary:
  - anaerobes: metronidazole
  - S. milleri: benzylpenicillin
  - coliforms: cefotaxime
  - consider co-amoxiclav or clindamycin, as often polymicrobial
- Mycobacterium tuberculosis: isoniazid, rifampicin, ethambutol and pyrazinamide.

Adult community-acquired pneumonia

Major presenting features: cough, fever, shortness of breath, chest pain.

The aetiological agent cannot be accurately predicted from clinical features.

Causes

- STREPTOCOCCUS PNEUMONIAE, Staphylococcus aureus, Haemophilus influenzae, S. pyogenes, K. pneumoniae
- MYCOPLASMA PNEUMONIAE, LEGIONELLA PNEUMOPHILA, Coxiella burnetii, Chlamydia psittaci, Chlamydia pneumoniae ('atypical' pneumonia agents)
- Moraxella catarrhalis, Neisseria meningitidis, Yersinia pestis, Chlamydia trachomatis (neonates), other Legionella spp.

Distinguishing features

See Table 5.2.

Complications

Pleural effusion, respiratory failure, adult respiratory distress sydrome (ARDS) and multi-organ failure are complications occurring with all the respiratory pathogens.

Streptococcus pneumoniae: empyema, bacteraemia, pericarditis, meningitis.

Staphylococcus aureus: empyema, pneumothorax, metastatic abscesses, bacteraemia, pericarditis, meningitis, lung abscess, toxic shock syndrome.

Mycoplasma pneumoniae: bullous myringitis, haemolysis, encephalitis.

Legionella pneumophila: empyema, lung fibrosis, neuropathies.

Coxiella burnetii: hepatitis, endocarditis.

Chlamydia psittaci: hepatitis, encephalopathy, myocarditis.

Investigations

- FBCt with differential WCCt
- Biochemical profile (liver function tests (LFTs), albumin and urea/creatinine)
- cold agglutinins (Mycoplasma pneumoniae)
- CXR
- Arterial gases or pulse oximetry
- Sputum:
  - Gram stain
  - antigen or DNA detection
  - culture (including Legionella) and sensitivities
- Blood culture (S. pneumoniae septicaemia occurs in 25%)
- Other samples that can be obtained for microscopy and culture: pleural fluid, tracheal aspirate, bronchoalveolar lavage (BAL) fluid, percutaneous aspiration
- Antigen detection:
  - S. pneumoniae (serum, sputum and urine)
  - L. pneumophila (sputum, urine)
- Acute and convalescent serum: all atypical pathogens
- CT scan thorax with bronchoscopy and/or percutaneous lung biopsy if persistent clinical and radiological disease.

Differential diagnoses

- Rarely, causes of chronic pneumonia can present acutely (see below)
- Pulmonary infarction, pulmonary oedema, allergic or cryptogenic alveolitis, Wegener's granulomatosis
- Collapse associated with foreign body, lung cancer.

Treatment

Streptococcus pneumoniae

- Intravenous (IV) benzylpenicillin (penicillin sensitive)
- IV cefotaxime (penicillin intermediate resistant)
- IV cefotaxime and IV vancomycin (penicillin resistant).

'Atypical' pneumonia

- Oral or IV clarithromycin (or erythromycin)
- IV rifampicin or IV ciprofloxacin in addition for severe Legionnaires' disease

	Pneumococcal pneumonia	'Atypical' pneumonia
Patient	Older, comorbidity	Younger, previously fit
History	Sometimes a preceding viral illness	Contact with birds, animals, affected family or water May be seasonal or outbreak
Incubation period	1–3 days	Usually > 7 days
Onset	Sudden with rigor	May be gradual
Influenza-like prodrome	Brief or absent	Present
Pleuritic chest pain	Usually present	Often absent
Cough	Starts early, < 2 days	Starts late, >5 days
Sputum	Initially dry, then rusty coloured	Initially dry, then mucopurulent
Non-respiratory symptoms	Uncommon	Common
Splenomegaly	Rare	Occasional
Rash	Rare	15% with *M. pneumoniae*
Total white cell count (× 10⁹/L)	Usually > 15	Usually ≤ 15
Hyponatraemia	Occasional	Marker for *L. pneumophila*
Hepatitis	Rare	Occasional
Chest X-ray	Well defined, lobar or lobular, unilateral (Fig. 5.2)	Patchy, poorly defined, bilateral
Sputum Gram stain	Gram-positive cocci	No organisms
Response to β-lactam antibiotics	Excellent*, rapid	Nil
Response to erythromycin	Excellent*, rapid	Excellent, slow
Main diagnostic method	Microscopy, culture or antigen detection	Serology, antigen detection for *L. pneumophila*

* Depending on sensitivity of isolate.

Table 5.2 Distinguishing features of adult community-acquired pneumonia.

- Oral or IV tetracycline is an alternative for adult patients if not *Legionella*.

Severe community-acquired pneumonia, aetiology unknown
- IV clarithromycin and IV cefotaxime (assuming high-level penicillin resistance rare in community isolates).

Prevention
Streptococcus pneumoniae
- Vaccination for high-risk groups

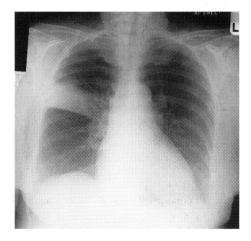

Fig 5.2 Right upper lobe consolidation in a patient with pneumococcal pneumonia.

• Antibiotic prophylaxis (children with sickle cell anaemia, consider for postsplenectomy and functional hyposplenism).

Chronic pneumonia

Major presenting features: *chronic cough, fever, night sweats, weight loss.*

Causes

• MYCOBACTERIUM TUBERCULOSIS, M. bovis, M. africanum
• Mixed anaerobic/aerobic bacteria
• Atypical mycobacteria, *Nocardia, P. pseudomallei, Actinomyces*
• *Cryptococcus neoformans, Aspergillus,* histoplasmosis, coccidioidomycosis
• *Paragonimus westermani.*

Distinguishing features

See Table 5.3.

Complications

Mycobacterium tuberculosis: respiratory failure, massive haemoptysis, lung fibrosis, pleural effusion/empyema, meningitis, focal extrapulmonary disease
Acute community-acquired pneumonia: see p. 44.

Investigations

Chronic pneumonia
• FBCt, differential WCCt, ESR
• Biochemical profile (LFTs, albumin, urea/creatinine)
• Pulse oximetry: arterial gases may be indicated
• CXR
• Sputum microscopy:
  • ZN and/or auramine
  • Gram
  • fungal
• Sputum culture:
  • Lowenstein–Jensen, Bactec (for tuberculosis)
  • fungal
  • anaerobic (*Actinomyces*)
  • prolonged standard (*Nocardia* and *Actinomyces*)
• If ZN is negative on three samples but appearances are consistent with tuberculosis, consider:
  • induced sputum
  • bronchoscopy and BAL

Table 5.3 Distinguishing features of cough and fever in adults.

	Tuberculosis	Acute community -acquired pneumonia
Epidemiology	Old, immigrant	Any age, any ethnic group
History	Recent or remote contact with tuberculosis	Preceding viral illness or contact with birds, animals, water or affected family
Onset	Subacute/chronic	Acute
Fever	Low grade	High and hectic
Weight loss	Moderate	Minimal
Haemoptysis	Not uncommon	Occasional
White cell count	Normal	Raised in S. pneumoniae
Haemoglobin	Anaemia	Normal
Chest X-ray		
Cavitation	Usual in adults	Uncommon (seen in S. aureus and K. pneumoniae)
Apical lesion	Usual	Uncommon
Distribution	Commonly bilateral, patchy	Unilateral and lobar for S. pneumoniae
Miliary	Occasional	Never
Response to therapy for acute pneumonia	Nil	Excellent

- gastric washings
- percutaneous lung aspiration/biopsy
- polymerase chain reaction (PCR) on specimens
- Serum: cryptococcal antigen.

Differential diagnoses
- Neoplasia, sarcoidosis, Wegener's granulomatosis, pulmonary embolus.

Treatment
Tuberculosis
- Isoniazid (6 months), rifampicin (6 months), pyrazinamide (2 months) and ethambutol (2 months).

Prevention
- Bacille Calmette–Guérin (BCG) vaccine
- Notification, screening of close contacts and prophylaxis of those persons suspected of having, or at risk of getting, primary subclinical infection (see under 'Tuberculosis')

Bronchiolitis and childhood pneumonia
Major presenting features: *cough, breathlessness, fever.*

Causes
Bronchiolitis
- RESPIRATORY SYNCYTIAL VIRUS (RSV)
- Parainfluenza, influenza virus, adenovirus, rhinovirus, *M. pneumoniae.*

Childhood pneumonia
- RSV, measles, varicella zoster, parainfluenza, influenza, adenovirus, coxsackie virus
- *Streptococcus pneumoniae, H. influenzae* (type b), *S. aureus, M. pneumoniae*
- *Mycobacterium tuberculosis.*

Distinguishing features
See Table 5.4.

Complications
Bronchiolitis: respiratory failure, apnoeic attacks, secondary bacterial pneumonia, recurrent wheezing, bronchiolitis obliterans.

Bacterial pneumonia: abscesses, cavities, pneumatocoeles (*S. aureus*), pleural effusion, empyema, bacteraemia, metastatic abscess, meningitis.

Diagnosis
- FBCt, differential WCCt
- Biochemical profile (LFTs, albumin, urea/creatinine)
- CXR
- Pulse oximetry or arterial gases
- Blood culture
- Throat swabs for viral and bacterial culture
- Nasal swab for viral culture
- Nasopharyngeal aspirate for RSV and other viral antigens
- Serum:
  - acute and convalescent sera (atypical agents)
  - pneumococcal antigen detection
- Urine: pneumococcal antigen detection.

Differential diagnoses
- Asthma, heart failure, congenital heart disease (left–right shunt), metabolic acidosis.

Treatment
Bronchiolitis
- Humidified oxygen, oral nutrition
- Nebulized ribavirin in those likely to develop severe disease (infants < 2 months, those born prematurely or with chronic cardio-respiratory disease, and immunocompromised children)

Bacterial pneumonia
- Humidified oxygen
- IV cefotaxime alone, or with, IV clarithromycin (if atypical agent possible).

Prevention
- *Haemophilus influenzae* type b (Hib) vaccine: conjugated pneumococcal vaccine where indicated (see later)
- Antibiotic prophylaxis with rifampicin for all close family contacts of index case of *H. influenzae* type b disease where there are other children aged < 5 years in the family

	Bronchiolitis	Bacterial pneumonia in childhood
Epidemiology		
Season	Winter, early spring epidemics	Commoner in winter, can occur any time
Age	Infants	Usually < 5 years
Prodromal upper respiratory tract symptoms	Present	Often absent
Fever	Low grade	High
Toxicity	Usually mild	Marked
Wheeze	Present	Usually absent
Recession	Present	Absent
Auscultation	Diffuse wheeze with crackles	Localized crackles or consolidation
White cell count	Normal or slightly raised	Raised
White cell count differential	Normal/lymphocytic	Neutrophilic
Chest X-ray		
Hyperinflation	Present	Absent
Abscesses or pleural effusion	Absent	May be present
Blood culture	Negative	May be positive

Table 5.4 Distinguishing features of cough and fever in infancy.

• RSV hyperimmune globulin may protect those at high risk of severe RSV infection.

Specific infections

Influenza
Epidemiology
Influenza is a common, highly infectious epidemic disease. It belongs to the myxovirus group and exists in two main forms, A and B. Influenza A is responsible for periodic worldwide epidemics (every 1–3 years) and unpredictable pandemics (every 1–2 decades). It also infects many other animals (including birds, pigs and horses) and shows much greater antigenic variation than B. Influenza C is uncommon and a cause of upper respiratory tract infection.

• Four pandemics of influenza A have been recorded (1918, 1957, 1968 and 1977): the first of these killed 200 000 persons in England and Wales, and 20 million worldwide
• There is a higher incidence of influenza in the winter months, with epidemics peaking in December/January. However, it can peak at different times in different winters
• Infectivity is very high and extends from shortly before symptoms start until shortly after the pyrexia settles
• The common occurrence of subclinical infections, and the short incubation period (IP), contribute to the rapid spread of disease
• The mode of transmission is by droplet infection and by hands and recently contaminated articles (e.g. handkerchiefs)
• The highest mortality is amongst the elderly
• Outbreaks in the Southern hemisphere from May to September usually predict the type of virus appearing in the Northern hemisphere the following winter
• Overall attack rates are 10–20% during an epidemic

• Influenza B causes sporadic infections and local epidemics about every 2nd year. The disease is usually mild and affects children particularly
• The IP is 1–3 days (range 12 h to 5 days).

Pathology and pathogenesis

Influenza A has the capacity to develop new antigenic variants at irregular intervals.

• The virus contains two surface antigens, namely a haemagglutinin (H antigen) and a neuraminidase (N antigen)
• These may change, and this is the basis of classification, together with the site and year of isolation (e.g. A/USSR/77 H1N1)
• Immunity develops specifically to H and N antigens; a change in either antigen will result in a loss of previous immunity
• Major change (antigenic shift) gives rise to pandemics and results from genetic reassortment between humans and the animal reservoir and the appearance of a novel H or N antigen (or both)
• Minor change (antigenic drift) gives rise to epidemics and results from the accumulation of random point mutations in the RNA and subsequent slight alteration of the H and/or N antigen
• Antigenic variation occurs less frequently with influenza B
• Three subtypes of influenza A are currently circulating: H1N1, H1N2 and H3N2
• The virus is cytopathic to the respiratory tract.

In pneumonitis, cytopathic involvement of the ciliated columnar epithelium of the trachea and bronchi, with sloughing of the epithelium, is found progressing to diffuse haemorrhage and hyaline membrane formation in fatal cases; inflammatory infiltrate is minimal.

Clinical features

The illness starts abruptly with fever, rigors, headache, aching eyes, myalgia, sore throat and dry cough. Most symptoms settle after 2–5 days but the cough and malaise may persist for 1–2 weeks. Physical examination shows only pyrexia and reddening of the fauces. In epidemics, mild attacks with short-lived fever and less prominent symptoms are very common.

Complications

Respiratory

• Pneumonitis (3%)
• Laryngotracheobronchitis and bronchiolitis
• Otitis media, secondary bacterial pneumonia
• Exacerbation of chronic respiratory disease
• Postviral fatigue syndrome.

Influenza pneumonitis occurs in 3% of cases. It is more likely in the elderly, in pregnant women, in those with compromised immunity and in patients with underlying cardiac or pulmonary disease. Development of marked dyspnoea, dry cough and cyanosis is associated with hypoxia and diffuse fine interstitial shadows on CXR. In young children, influenza A or B may cause laryngotracheobronchitis or bronchiolitis. Secondary bacterial pneumonia (*S. aureus*, *S. pneumoniae* or *H. influenzae*) is an important cause of death. Ten per cent of adults with confirmed influenza infection admitted to hospital have coincident *S. aureus* infection. It is usually seen in the elderly and those with underlying respiratory disease. CXR reveals multiple areas of consolidation.

Non-respiratory

• Guillain–Barré syndrome, myelitis, encephalitis
• Reye's syndrome
• Myositis and myoglobinuria
• Myocarditis and pericarditis.

Diagnosis

In the presence of a confirmed outbreak, a patient with influenza-like symptoms is likely to have influenza. Confirmation can be achieved by demonstrating the following:

• virus isolation from nose or throat swabs
• a fourfold rise or fall in antibody titre, or a single high titre by complement fixation or haemagglutination assay
• antigen detection on pharyngeal aspirates
• specific nucleic acid (NA) amplification by PCR.

If secondary infection is suspected:

• blood culture and sputum Gram stain and culture should be performed.

Treatment
The severity of influenza is variable, but most patients require only symptomatic therapy.
• Amantadine will shorten the duration and reduce the severity of symptoms of influenza A by one-third if started within 48 h
• Zanamivir and oseltamivir are neuraminidase inhibitors and are active against influenza A and B. They reduce the amount of virus that is shed and shorten the duration of symptoms if given within 48 h of symptom onset
• Treatment for influenza should be reserved for persons at increased risk of developing severe complications and only when influenza is circulating in the community
• Antibiotics (e.g. cefotaxime) with activity against the likely secondary pathogens (*S. aureus*, *S. pneumoniae* and *H. influenzae*) should be used in all ill patients with respiratory complications
• Aspirin should not be used in children because of the risk of Reye's syndrome.

Prevention
• Global monitoring of influenza activity allows forewarning of impending epidemics. Early detection of influenza-like illness in the UK is done through a network of 'spotter' general practitioners
• Influenza vaccine is recommended for persons with chronic respiratory, cardiac or renal disease, diabetes or other endocrine disorders, for those with immunosuppression (including HIV), for old institutionalized persons, and for health-care workers involved in patient care
• Amantadine or rimantadine will protect 70% from symptomatic illness and should be considered for unvaccinated exposed persons who are health-care workers, those at high risk for complicated disease, or those in whom immunization is contraindicated. Similarly, both neuraminidase inhibitors have been shown to be effective in preventing symptomatic illness, although experience is less.

Prognosis
The death rate increases with age, underlying illness and the presence of complications, particularly pneumonitis. The 1989–1990 epidemic was thought to be responsible for 29 000 excess deaths in England and Wales.

Bronchiolitis
Epidemiology
Bronchiolitis is an acute lower respiratory tract infection (RTI) of infancy characterized by wheezing and hyperexpansion and resulting from inflamed, obstructed small airways. RSV is a major cause (60% of cases), as are the para-influenza viruses (20%). Occasionally, influenza virus, adenovirus, rhinovirus, *M. pneumoniae* and enterovirus are responsible. RSV also causes tracheobronchitis, pneumonia and upper RTI with otitis media; asymptomatic infection is uncommon.
• RSV is an RNA paramyxovirus with two major subtypes, A and B
• Immunity is incomplete and repeated infections are common; severe disease rarely occurs after primary infection. By 2 years of age nearly all children have serological evidence of infection
• Epidemics occur annually in winter and early spring
• Disease is commoner in boys, breastfed babies and those from lower socioeconomic groups
• The most severe disease occurs in the youngest infants (<2 months old), those born prematurely or with chronic cardiorespiratory disease, and the immunocompromised
• Spread is by droplet infection to mucous membranes, either airborne or by direct contact; viral shedding lasts 1 week
• The virus is readily inactivated by disinfectants and survives only a few hours on surfaces
• RSV can also cause severe pneumonitis in the elderly
• The IP is 3–5 days (range 2–8 days).

Pathology and pathogenesis
RSV causes patchy peribronchial inflammation with oedema, necrotic material and fibrin pro-

duction leading to obstruction. If partial, this leads to air trapping and hyperinflation; if complete, it leads to collapse. Interstitial pneumonia may also be present. Local immune response is probably important in pathogenesis as shown by raised specific immunoglobulin E (IgE) antibodies in nasopharyngeal secretions. However, systemic antibody is poorly protective because disease may occur despite maternal or vaccine-stimulated antibody. An inactivated candidate vaccine resulted in an exaggerated response to wild virus infection with more severe disease.

Clinical features
The characteristics of RSV bronchiolitis are as follows:
• fever, coryza and cough precede the development of respiratory distress by 1–2 days
• pallor, tachypnoea, tachycardia and restlessness supervene
• supraclavicular, intercostal and subcostal recession of the soft tissues on inspiration develop, associated with expiratory wheeze
• auscultation reveals wheeze with or without diffuse crackles
• cyanosis, aggravated by coughing or by attempts at feeding, develops in severe cases
• fever is intermittent and rarely exceeds 39°C
• the disease lasts 3–7 days with gradual recovery over 1–2 weeks.

Complications
• In primary RSV, 1% require hospitalization
• The major complication is respiratory failure, which is rare in previously healthy children. Up to two-thirds of fatal cases occur in patients with cardiopulmonary disease or who are immunosuppressed. Apnoea and hypoxia are not uncommon in hospitalized infants, whereas secondary bacterial infection is unusual. Hyperreactive airways and asthma may be linked with bronchiolitis in infancy. Bronchiolitis obliterans is a very rare and severe complication which appears to follow adenovirus bronchiolitis.

Diagnosis
The diagnosis is primarily epidemiological and clinical although the condition must be distin-

guished from asthma, which is usually recurrent, and from pneumonia, where the striking signs of airway obstruction are not normally seen. Non-specific features supporting RSV include:
• normal or slightly raised WCCt with lymphocytic differential
• hyperinflation, prominent bronchial wall markings and multiple areas of collapse-consolidation on CXR
• hypoxaemia without hypercapnia (unless respiratory failure is developing).
Specific diagnosis is made through one of the following:
• virus isolation from a nose or throat swab
• antigen or RNA detection from nasopharyngeal aspirates
• a fourfold rise or fall in antibody titre or a single high titre.

Treatment
Children with signs of lower respiratory tract involvement are best treated in hospital in isolation. Severe cases should be:
• nursed in oxygen with monitoring of Po_2 or oxygen saturation
• nursed in a humid atmosphere
• tube fed if the infant is distressed by normal feeding
• sucked out repeatedly, which helps to maintain the airway
• carefully observed for overhydration which must be avoided
• considered for ventilation and aerosolized ribavirin if the disease is particularly severe.
Antibiotics, bronchodilators and corticosteroids are ineffective.

Prevention
• Control of cross-infection in children's wards
• Infants should not be unnecessarily exposed to adults and older siblings who are suffering from upper RTI
• RSV-IVIG (immunoglobulin with high titres of neutralizing antibody to RSV) or palivizumab (a humanized murine monoclonal antibody) can be given to protect persons at high risk of severe disease during outbreaks
• Trials with killed RSV vaccine were dis-

continued in view of the occurrence of severe clinical disease in the vaccinated.

Prognosis

Deaths are exceptionally rare unless the infant has underlying disease, was born prematurely or is < 2 months old. In these categories there is an appreciable mortality (5–35%).

Chlamydia pneumoniae infection

Epidemiology

Chlamydia pneumoniae is the most recently identified and probably the commonest chlamydial pathogen of humans.

- It is a primary human pathogen without an animal reservoir
- Transmission is by aerosol
- Men are affected more frequently than women
- Antibodies are rare in children; by 20 years of age, 50% have antibodies
- Most infections are mild, subclinical infections are common and reinfections throughout life occur
- Pneumonia is the commonest manifestation of clinical disease, but bronchitis, 'influenza-like' illness, pharyngitis, sinusitis and otitis media also occur
- It is commonly associated with outbreaks in the community and amongst closed groups, and may be nosocomial
- It causes approximately 10–15% of community-acquired pneumonias
- it is often coidentified with other pathogens (e.g. *S. pneumoniae*)
- The IP is 10 days.

Pathology and pathogenesis

The organisms are small, coccoid bacteria which are obligate intracellular parasites. The pathogenesis is presumed to be very similar to that for *C. psittaci* (see p. 53).

Clinical features

Clinical distinction of pneumonia caused by *C. pneumoniae* from other aetiologies of pneumonia is difficult. Nevertheless, the following features are characteristic:

- pharyngitis and hoarseness
- a biphasic illness with initial improvement of upper respiratory tract symptoms
- the development of pneumonia 1–3 weeks later, with acute onset of dry cough and fever without chest pain
- a mild illness
- a relapsing course with persistent cough.

Complications

Severe disease leading to respiratory failure may occur in the elderly or those with underlying chest or heart disease. Chronic sinusitis sometimes occurs.

Diagnosis

The CXR usually has a single infiltrate, the WCCt is normal and the ESR is raised. Laboratory confirmation is by type-specific immunofluorescence assay. The standard chlamydial antibody test is the complement fixation text. This cannot differentiate *C. pneumoniae* from *C. psittaci* and lacks sensitivity, especially for reinfections. Distinction from *C. psittaci* requires a type-specific immunofluorescence assay. Antibody rises can be slow and blunted by prompt antibiotic therapy.

Treatment

Tetracycline and erythromycin are both effective against *C. pneumoniae*. The newer quinolones (e.g. ciprofloxacin) and macrolides (e.g. clarithromycin) may be effective alternatives. Treatment should be continued for 2 weeks. Evidence that antibiotics alter the course of the disease is lacking.

Prevention

No effective vaccine is available.

Prognosis

The illness is benign. The very rare fatality results from respiratory failure.

Psittacosis

Epidemiology

Psittacine birds are an important source of human infection with *C. psittaci*, but other birds including turkeys, ducks and pigeons may also cause infection. The term ornithosis is therefore more correct.

• Infection is more common in adults, reflecting greater exposure to infected birds
• Transmission is through handling infected ill birds or, less commonly, by inhaling organisms in dried dust from bird droppings. Outbreaks occur rarely and usually in relation to infected sources at work
• Transmission between birds, especially when held captive, is common. Hence, it is often recently imported birds that are the source of infection
• Only 20% of patients with psittacosis give a history of bird contact
• Many infections are mild or subclinical
• Birds with *C. psittaci* infection are unwell, although sometimes the features may only indicate a mild illness
• *Chlamydia psittaci* accounts for < 3% of cases of community-acquired pneumonia in the UK; there are approximately 300 cases annually
• Human to human transmission may occur but is rare
• The IP is 10 days (range 7–15 days).

Pathology and pathogenesis

The primary site of disease in humans is the lung. The bacterium enters through the respiratory tract and, after a transient bacteraemia, seeds the reticuloendothelial system. In the lung, there is an alveolitis followed by an interstitial exudate. Mucus plugging is a feature of chronic cases. A second bacteraemia coincides with the onset of symptoms. Hepatitis is often observed on biochemical testing, and rarely may be frank and the primary presentation; biopsy shows reactive hepatitis but may also show granulomata.

Clinical features

Characteristics are as follows:
• the onset is abrupt, with fever, rigors, headache, myalgia and a dry cough (occasionally with blood streaking); pleuritic pain is rare
• clinical examination of the chest may be unremarkable despite marked involvement on CXR

• splenomegaly is found in one-third and a rash is rarely observed (Horder's spots)
• abdominal pain, diarrhoea, jaundice, pharyngitis and a mild encephalopathy are occasionally seen
• there is a relative bradycardia.

Complications

• Respiratory failure
• Hepatitis
• Encephalopathy, glomerulonephritis, myocarditis, endocarditis.

Diagnosis

Psittacosis is suggested by:
• history of bird exposure
• a CXR showing soft, patchy infiltrates radiating out from the hila and involving the lower zones
• a normal WCCt and ESR
• mildly deranged LFTs.
Psittacosis is confirmed by:
• a fourfold rise or fall in antibodies or a single high titre by complement fixation. Distinction from *C. pneumoniae* requires a type-specific immunofluorescence assay. Prompt treatment can delay and/or blunt the antibody response.

Treatment

The treatment of choice is tetracycline; erythromycin is also effective and is the first-line agent in children. Therapy should be continued for 2 weeks. Rifampicin, ciprofloxacin and the new macrolides (e.g. clarithromycin) have also been used successfully but clinical experience is less.

Prevention

• Quarantining and prophylactic administration of tetracycline-medicated seed to imported birds
• Tetracycline supplemented to animal feeds can prevent chlamydial infections in poultry flocks but may create wider problems with antibiotic-resistant *Salmonella*
• There is no vaccine.

Prognosis

Mortality is 1% of cases, but in severe cases complicated by renal insufficiency it may rise to 20%.

Q fever

Epidemiology

Q fever is caused by the rickettsia *C. burnetii*. It is a worldwide zoonosis affecting wild animals as well as domestic livestock. Infection is endemic in cattle, sheep and goats, where most human infection is acquired.

• Infection is largely subclinical in animals
• Organisms are excreted in milk, urine and faeces of infected animals and in high concentration in amniotic fluid and placenta. They are resistant to heat, desiccation and many disinfectants
• Transmission is by aerosols or environments (e.g. dust) contaminated by parturition products. Rarely, contaminated raw milk may be responsible
• Few organisms are required to establish infection
• It is most common between April and June, possibly related to the calving and lambing season
• It is the cause of < 1% of community-acquired pneumonia, and an occasional cause of acute hepatitis
• It is more common as a cause of community-acquired pneumonia in north-west Spain and Canada than in the UK
• It is an occupational hazard for abattoir workers, veterinarians, etc.; outbreaks have occasionally been reported in these groups and in others exposed to dust from animals
• A history of occupational exposure is only found in 5–10%
• The IP is 14–28 days.

Pathology and pathogenesis

After inhalation, the organism multiplies in the lung, followed by bacteraemia, which coincides with the onset of symptoms. A lymphocytic interstitial pneumonitis occurs, often accompanied by a mild hepatitis. Liver (and bone marrow) biopsy may show 'doughnut' granulomata with a fatty fibromatous centre. Rarely, patients may present with chronic infection,

usually affecting the aortic valve, and with typical clinical and pathological features of endocarditis. Chronic liver disease leading to cirrhosis has also been described.

Clinical features

In 50% the illness is subclinical.
• The illness starts abruptly with fever, headache, myalgia and occasionally cough
• Clinical or radiological evidence of pneumonia develops in 30% of patients. Clinical signs may be few despite quite extensive X-ray abnormalities
• Hepatosplenomegaly is found in 50% of cases
• Fever usually continues for 1–2 weeks.

Complications

• Hepatitis
• Chronic infection: endocarditis, cirrhosis (acute infection nearly always unrecognized). This may occur from between 1 and 20 years after the acute illness.

Diagnosis

• Thrombocytopenia is common
• Demonstration of a fourfold rise or fall in or single high titre of phase 2 antibodies by complement fixation. Phase 1 antibody is absent in acute infection
• Chronic Q fever should be suspected in all culture-negative cases of endocarditis; both phase 1 and 2 antibodies are detectable
• Phase 1 and 2 antibodies have been found in individuals with no evidence of cardiac involvement, and it is likely in these persons that chronic liver infection is present
• Antigen and DNA detection is possible in tissue samples.

Treatment

The treatment of choice is a tetracycline (doxycycline is commonly used), but response may be slow. Erythromycin, ciprofloxacin, rifampicin and the newer macrolides (e.g. clarithromycin) are also active. Treatment of endocarditis requires prolonged therapy (2–3 years, sometimes indefinitely), preferably with a combination of antibiotics (e.g. doxycycline and a quinolone or co-trimoxazole and rifampicin); surgery may need to be considered. Management of endocarditis is dealt with in more detail in Chapter 8.

Prevention
- Appropriate disposal of parturition products
- Pasteurization of raw milk
- An unlicensed vaccine is available, although severe reactions may occur in previously exposed individuals

Prognosis
In acute Q fever, fatalities are exceptionally rare. In Q fever endocarditis, mortality is 10–15% of cases.

Mycoplasma pneumonia
Epidemiology
Mycoplasma pneumoniae is a common pathogen of worldwide distribution resulting in both upper (rhinitis, pharyngitis, myringitis, conjunctivitis) and lower (pneumonia, tracheobronchitis) RTI. Infection is usually mild.
- It accounts for approximately 10% of community-acquired pneumonias in the UK; during outbreaks this may treble
- Infection rates are greatest in children and young adults
- Transmission is by droplets and requires close contact. Spread is slow, with weeks between cases, but attack rates at home are high (75% children and 35% adults) with a slight preponderance in males
- Pneumonia is the most commonly recognized form of infection and affects mainly older children and young adults. However, only 5% of infected persons develop pneumonia, nondescript upper RTI being the commonest manifestation, especially in infants and young children
- Infection is endemic in large communities without evidence of seasonal prevalence. However, periodic outbreaks in autumn and early winter occur every 3–4 years
- The IP is 6–23 days.

Pathology and pathogenesis
Attachment to respiratory tract epithelium is by P1 protein binding to neuraminic acid glycoprotein receptors. This is sequentially followed by ciliostasis, deciliation and cytopathic effects. The organism also stimulates both T- and B-cell lymphocyte responses; the relative importance of direct mycoplasmal damage, as opposed to immune mechanisms, has not been determined. I and i erythrocyte antigens serve as receptors for *M. pneumoniae* resulting in cold haemagglutinins and the complication of intravascular haemolysis. Fatal cases show patchy alveolar pneumonitis and peribronchial infiltrate of lymphocytes and plasma cells. Immunity follows primary infection but occasional second infections have been described.

Clinical features
Characteristic features include:
- a 2- to 4-day prodrome of gradual-onset high fever, rigors, malaise, myalgia and headache
- upper RTI (coryza, tonsillopharyngitis, myringitis or conjunctivitis) in 25–50%
- development of a dry cough which is later productive (33%) of mucopurulent or occasionally blood-tinged sputum
- absence of pleuritic chest pain
- unremarkable chest examination despite marked radiological features
- extrapulmonary manifestations: skin rashes (maculopapular, urticarial, erythema multiforme, erythema nodosum), diarrhoea and vomiting, and arthritis each occur in approximately 15–25% of cases.

Complications
Pulmonary
- Bullous myringitis (visible bleb formation on the tympanic membrane)
- Respiratory failure, adult respiratory distress syndrome (ARDS)
- Pleural effusion (15%)
- Cavitation, pneumatocoeles, bronchiectasis, pulmonary fibrosis (all rare).

Extrapulmonary
- Intravascular haemolysis

- Meningoencephalitis, transverse myelitis, and cranial and peripheral neuropathies (all rare)
- Hepatitis, glomerulonephritis, pancreatitis, myocarditis and pericarditis (all rare).

Diagnosis

Non-specific features supporting *M. pneumoniae* include:
- presence of current or antecedent upper RTI and extrapulmonary features
- patchy consolidation of one or other lower lobes (80%), often bilateral (30%). Hilar lymphadenopathy may be present
- presence of cold haemagglutinins (50%)
- total WCCt of $< 10 \times 10^9$/L (90%)
- mildly deranged LFTs.

Specific diagnosis is made through one of the following:
- isolation of *M. pneumoniae* from throat or sputum culture using selective liquid media
- detection of nucleic acid in tissue or sputum using DNA probes
- fourfold rise or fall in or a single high antibody titre by complement fixation or enzyme-linked immunoadsorbent assay (ELISA).

Treatment

Both erythromycin (or the newer macrolides) and tetracycline are effective. Erythromycin or clarithromycin tend to be preferred because of the availability of parenteral formulations, the better activity against *Legionella* and the problem with tetracycline of staining children's teeth. Therapy should be for 2 weeks. The clinical response is not rapid.

Prevention

- There is no vaccine for mycoplasmal infection. Isolation in hospital is advised
- Prophylactic azithromycin substantially reduces the secondary attack rate in institutions.

Prognosis

Mycoplasma pneumoniae infection is usually mild and self-limiting. Where pneumonia occurs it tends to be a benign illness. During outbreaks, occasional fatalities are always reported but this is a rare event.

Legionnaires' disease

Epidemiology

Legionella pneumophila was first identified in 1976 and was subsequently identified as the cause of several earlier outbreaks of pneumonia (Legionnaires' disease) and influenza-like illness (Pontiac fever). Over 30 other species of *Legionella* have been identified, of which 16 can cause a similar respiratory illness which is termed legionellosis.

Legionella pneumophila
- A widespread, naturally occurring aquatic organism. Survival has been associated with blue-green algae and free-living acanthamoeba
- Transmitted via airborne water droplets; possible sources include air-conditioning cooling towers, hot-water systems, humidifiers and showers. Case-to-case transmission does not occur
- Responsible for 5% of community-acquired pneumonias; there are approximately 250 cases annually in the UK
- Divided into 14 serogroups, of which serogroup 1 is most important
- Very occasionally responsible for infection at other sites (wound infection, endocarditis).

Legionnaires' disease
- Is most frequently seen in middle-aged males and in September/October in the UK
- Results in subclinical infection infrequently, and infection in childhood very rarely
- May occur sporadically or in outbreaks, often centred on institutions such as hotels or hospitals. The attack rate is <5% in outbreaks. Clusters of cases are often linked to Mediterranean resorts; however, only 23% of cases occur in clusters
- 52% of UK cases are related to travel, 91% of these relate to travel abroad
- Most common in patients with diabetes, with malignancy, undergoing dialysis or on immunosuppressives, and in heavy smokers and drinkers
- Has an IP of 2–10 days.

Pontiac fever
- Is caused by a hypersensitivity reaction to *L.*

pneumophila. The organism is never isolated from cases and diagnosis is by serology
• Has an attack rate of 90% in outbreaks
• Has an extremely good prognosis without specific therapy
• Has an IP of 1–2 days.

Pathology and pathogenesis

Legionella pneumophila causes alveolar and interstitial inflammation with the development of patchy or widespread consolidation in the lung. It is an intracellular pathogen, parasitizing macrophages where it multiplies and evades host defences. Successful resolution of infection depends primarily upon cell-mediated immunity.

Clinical features

Distinction from other causes of community-acquired pneumonia is difficult. In favour of Legionnaires' disease are the following:
• severe influenza-like symptoms (80%) with fever, rigors, myalgia and headache
• prominence of extrapulmonary symptoms (35%), such as diarrhoea, abdominal pain, haematuria and confusion
• the later development of chest symptoms with dry cough and dyspnoea. Haemoptysis, chest pain and a productive cough may follow on later
• on examination, systemic toxicity, relative bradycardia, tachypnoea and crepitations on auscultation.

Complications

• Acute respiratory failure
• Pleural effusion/empyema (5%), cavitation (rare)
• Residual fibrosis
• Renal failure
• Cranial and peripheral nerve palsies, ataxia.

Diagnosis

Non-specific features supporting *L. pneumophila* include:
• exposure history to potential source
• limited chest signs with extensive involvement on CXR

• patchy consolidation, often bilateral (30%) and more frequent involvement of the lower zones
• hyponatraemia, hypophosphataemia, and modest elevations of transaminases, alkaline phosphatase, bilirubin, urea and creatinine
• total WCCt of < 15 × 10^9/L and lymphopenia
• microscopic haematuria and proteinuria (30%).
Specific diagnosis is made through one of the following:
• sputum culture (positive in 15%)
• antigen detection in urine (tissue or sputum can also be used)
• fourfold rise or fall in or single high antibody titre by immunofluorescence (IFAT) or agglutination (RMAT, Latex); 25% of patients only seroconvert 4–8 weeks after onset.

Treatment

• Clarithromycin is the antibiotic of choice (because of the IV formulation). In severe cases, either rifampicin or ciprofloxacin should be added. Treatment should be continued for 2 weeks
• Intensive care may be needed.

Prevention

With an established outbreak, it is important to identify the source or site of water exposure and, if possible, close it down and screen at-risk patients (e.g. those in a dialysis unit). Treatment of contaminated water supply by hyperchlorination or maintenance of water storage temperatures at > 60°C and outlet temperatures at > 50°C will reduce the risk of institution-associated infection.

Prognosis

The mortality rate in epidemics of *L. pneumophila* pneumonia is between 10 and 25% of cases.

Pneumococcal pneumonia

Epidemiology

Streptococcus pneumoniae is responsible for 20–30% of cases of community-acquired pneumonia (incidence 1 case/1000 adults in the UK), is the second most frequent pathogen in bacterial meningitis, and is a major aetiological agent in sinusitis, otitis media and infective exacerbations of chronic chest disease. Pneumococci are:

• commonly carried asymptomatically in children (30%) and adults (15%). Carriage is higher where there is overcrowding or a closed group
• transmitted by droplet infection; this is augmented by overcrowding and coexistent viral upper RTI. Pharyngeal colonization usually precedes pneumonia.

Pneumococcal pneumonia:
• is more frequent in patients with HIV infection, the nephrotic syndrome or chronic lung, heart and kidney disease; in splenic dysfunction or post splenectomy; in alcoholic liver disease or cirrhosis; in diabetics; and in patients with certain malignancies (e.g. myeloma, lymphoma) and severe immunosuppression
• is commoner in males, persons aged >40 years, during winter and in closed groups: in this context outbreaks occasionally occur
• is a common complication of respiratory viral infections, particularly influenza
• has an IP of 1–3 days.

Pathology and pathogenesis

Pneumococci are capsulate (imparting virulence) and the antigenic differences between the capsules form the basis of the serotypic classification. There are over 80 serotypes, and certain serotypes are more commonly associated with clinical disease; polyvalent pneumococcal vaccine contains 23 of the commonest serotypes isolated. Following inhalation, S. pneumoniae rapidly multiply in the alveolar spaces. Subsequently, the classical pathological progression of congestion, through red hepatization to grey hepatization and finally to resolution, occurs. The inflammation is purely alveolar and there is no development of necrosis or postpneumococcal fibrosis.

Clinical features

In acute pneumococcal pneumonia:
• the onset is abrupt, with a rigor followed by a high fever and general prostration
• a dry cough, which later becomes productive of rusty sputum, and chest pain develop in 75% of patients
• examination reveals toxicity, tachypnoea and features of consolidation
• labial herpes simplex is often present.

Complications

Local
• Pleural effusion (10%)
• Empyema (< 1%)
• Pericarditis.

Empyema should be suspected where persistent fever and leucocytosis occur, despite adequate antibiotic therapy; aspiration and/or drainage is indicated. Pericarditis results from contiguous spread.

Systemic
• Bacteraemia (occurs in 25% and doubles mortality)
• Meningitis, arthritis, endocarditis (rare).

In patients without a functioning spleen, a fulminant course may be observed, with death in 24h.

Diagnosis

Non-specific features supporting S. pneumoniae include:
• presence of chest pain and few extrapulmonary features
• leucocytosis $> 15 \times 10^9$/L
• radiological evidence of well-defined area(s) of consolidation (Fig. 5.2)
• sputum Gram stain showing Gram-positive diplococci.

Confirmation of infection may come from:
• isolation of a pure and heavy growth of S. pneumoniae from sputum culture
• recovery of S. pneumoniae from blood culture or pleural fluid
• pneumococcal antigen detection in sputum, pleural fluid, urine and/or serum.

Treatment

The recognition of penicillin-resistant pneumococci has altered management in many areas of

the world. These strains are characterized as follows:
• they are associated with prior antibiotic use, the extremes of age and hospital acquisition
• they account for >30% of all isolates from clinically important infections in certain areas of North and South America, Spain and South Africa
• they can be split into those with intermediate sensitivity to penicillin (high-dose benzylpenicillin effective except in meningitis) and fully resistant strains (benzylpenicillin completely ineffective)
• they may be associated with high-level resistance to other antibiotics.

Where penicillin-sensitive pneumococci have been isolated, the treatment is benzylpenicillin. Where intermediately resistant strains are not uncommon but fully resistant strains rare, cefotaxime should be used; where fully resistant strains are endemic, a combination of cefotaxime and vancomycin is indicated. When the sensitivity profile of the pneumococcus is known, treatment can be changed accordingly.

Prevention
• Immunization with pneumococcal vaccine containing capsular polysaccharide of the commonest 23 serotypes gives 60–70% protection. It is indicated for the at-risk groups mentioned above. Immunity is long-lasting. The currently available unconjugated vaccine is poorly immunogenic in children <2 years of age. A 7-valent protein conjugated vaccine has recently been licensed for children under 2 years of age at risk for invasive disease (e.g. with sickle cell disease).
• Chemoprophylaxis with oral penicillin is indicated in children with sickle cell anaemia and should be considered for patients with functional hyposplenism or post-splenectomy.

Prognosis
Pneumococcal pneumonia is a leading cause of death. The conditions that predispose to infection also increase the likelihood of severe and fatal disease.

In community-acquired pneumonia, the presence of two of the following parameters is associated with a 21-fold increased mortality:
• respiratory rate >30/min
• diastolic blood pressure <60 mmHg, systolic <90 mmHg
• blood urea >7 mmol/L
• confusion (AMT score ≤8).

Additional poor prognostic features are hypoxia (saturations <92% or $P_{a}O_{2}$ <8 kPa), bilateral or multiple lobe involvement, coexistent morbidity, and age over 50.

Staphylococcus aureus pneumonia
Epidemiology
• Causes <1% of cases of community-acquired pneumonia
• Often complicates influenza (40% of those with staphylococcal pneumonia admitted to hospital and 50% of those admitted to ICU have evidence of influenza infection)
• Is a not infrequent cause of hospital-acquired pneumonia, when it may be caused by methicillin-resistant S. aureus (MRSA)
• Is more frequent in the winter months.

Pathology and pathogenesis
Lung infection may result from necrotizing pneumonia of healthy lungs, septic embolus causing abscess formation (e.g. from injection drug user or right-sided endocarditis), or secondary infection of an infarct. Necrotizing pneumonia is often associated with suppuration, cavitation and abscess formation. Histology reveals intense neutrophilic infiltration, and microabscess formation with necrosis of lung parenchyma. Dissemination to other sites may occur.

Clinical features
• Presentation is with acute onset fevers, rigors and systemic toxicity. Cough, shortness of breath and chest pain are usually present and clinical examination typically reveals signs of consolidation, and/or a pleural rub
• Where infection is more insidious, clubbing may be seen
• Where the pneumonia has resulted in

necrosis of lung tissue, cavitation and abscess formation may occur: the cough may suddenly become purulent with rupture of the cavity into the bronchus.

Diagnosis

Non-specific features supporting *S. aureus* include:

- history of preceding influenza-like illness
- presence of chest pain and few extrapulmonary features
- leucocytosis $> 15 \times 10^9$/L
- radiological evidence of necrosis and cavity formation within the consolidation
- sputum Gram stain showing Gram-positive cocci in grape-like clusters.

Confirmation of infection may come from:

- isolation of a pure and heavy growth of *S. aureus* from sputum, blood or pleural fluid culture.

Differential diagnosis

Necrotizing pneumonia:

- *Klebsiella* (< 1% of community-acquired pneumonia, systemic toxicity, upper lobes, bulging fissure, high mortality)
- Aspiration pneumonia (after vomiting/unconsciousness; in achalasia, bulbar palsy, vocal cord paralysis, alcoholics; polymicrobial involving anaerobes, right lower lobe)
- *Fusobacterium necrophorum* (rare, preceding sore throat, bacteraemia and toxaemia, ectopic abscesses).

Lung abscess:

- Aspiration, amoebic, tuberculosis.

Complications

- Pneumothorax, empyema and pleural effusion
- Septicaemia, meningitis, pericarditis, endocarditis, cerebral abscess
- Respiratory and/or multiorgan failure
- Toxic shock syndrome
- Residual thin-walled 'pneumatocoeles' on recovery.

Treatment

- Oxygen and IV fluids as required
- Where *S. aureus* is suspected but not

confirmed and the infection has been acquired in the community: high-dose flucloxacillin and clarithromycin. A second antistaphylococcal drug may be indicated, dependent on the severity of the infection (e.g. gentamicin, rifampicin)

- In hospital-acquired infection, teicoplanin or vancomycin (to cover MRSA) should be used with or without a second antistaphylococcal drug (as above): additional antibiotics may be indicated to cover Gram-negative and anaerobic organisms depending on the situation.

Whooping cough

Epidemiology

Whooping cough is a highly infectious disease caused by *Bordetella pertussis*. The characteristic whoop is a result of a sharp indrawing of breath which follows a paroxysm of coughing. Several viruses (particularly adenovirus), *M. pneumoniae*, *C. trachomatis* and *B. parapertussis* can all produce a similar but milder illness. Whooping cough:

- is a major cause of childhood illness (600 000 deaths annually) worldwide. Most infections occur in unimmunized infants
- is a disease of young children; 50% of cases occur in those < 1 year of age, when the mortality is highest: there is no effective transmitted maternal immunity. Adults can get the disease, when the illness is milder; they may be the source for children
- is an endemic disease with epidemics every 3–5 years
- is readily transmissible from clinical cases by aerosol, with attack rates of over 50%
- occurs more frequently and seriously in females
- confers good immunity after a single attack; second attacks are rare
- is infectious from the onset of catarrhal symptoms for up to 1 month
- has an IP of 7–14 days.

Pathology and pathogenesis

The primary pathology is in the lungs, with systemic manifestations being secondary to toxaemia and respiratory complications. Pulmonary atelectasis is common and is due partly to bronchial blockage by viscid mucus and partly to bronchial and peribronchial inflammation. The extent of collapse varies from small subsegmental areas to a whole lobe. In fatal cases, the major pathological feature is haemorrhage, which can be explained by the raised pressure caused by paroxysmal coughing. *Bordetella pertussis* produces a number of biologically active substances:

• the filamentous haemagglutinin and the agglutinogens (important in attaching to ciliated respiratory epithelium)
• tracheal cytotoxin
• dermonecrotic and pertussis toxins (causing ciliostasis and local cell damage)
• pertussis and adenylate cyclase toxins (interfering with phagocyte function)
• pertussis toxin (causing the systemic manifestations of disease). This toxin has a typical A/B unit structure with binding and active toxin components and is directed against guanine nucleotide binding (G) proteins.

Clinical features

The early clinical features, with coryza, mild fever and dry cough, are non-specific but the child is highly infectious. However, over the next 1–2 weeks other signs will be seen.

• The cough becomes progressively more severe
• Paroxysms develop, with the characteristic whoop on indrawing of breath. The whoop is uncommon in infants and adults, and may be absent in those with partial immunity
• Cough and whoop become more frequent, especially at night, occurring up to 50 times/day and usually terminated by vomiting. Paroxysms can be triggered by handling or examining the child, or by exposure to cold air
• Cyanosis during paroxysms commonly occurs. Apnoeic attacks may be the only symptoms in the very young, as may chronic cough in the adult. After 2–4 weeks, the paroxysms become gradually less frequent, but may continue to occur, especially at night, for up to 6 months.

Any minor intercurrent infection may retrigger paroxysms.

Complications

Respiratory tract

• Pulmonary collapse (particularly the lower lobes and the right middle lobe)
• Secondary bacterial pneumonia.

Other complications

• The increased pressures associated with paroxysmal coughing can result in subconjunctival haemorrhages, epistaxis, facial and truncal petechiae, subcutaneous emphysema, pneumothorax, abdominal hernias and rectal prolapse. Ulceration of the lingual frenulum due to repeated tongue movement over the lower incisors occasionally occurs
• Convulsions complicate a minority (2%) and may result from cerebral haemorrhage (rare), anoxia or neurotoxic properties of pertussis toxin
• Vomiting can lead to weight loss and malnutrition.

Diagnosis

During the early catarrhal phase, the diagnosis can only be suspected from a history of contact. With the development of typical paroxysms and whooping, the diagnosis becomes clinical. Non-specific features supporting *B. pertussis* include:

• a persistent paroxysmal cough, worse at night, and accompanied by vomiting
• the appearance of the whoop
• leucocytosis ($>20 \times 10^9$/L) with lymphocytosis ($>80\%$) (not usually seen in infancy)
• pulmonary collapse or consolidation (20%).

Confirmation can be obtained from:

• per nasal swab culture on selective charcoal-based blood or Bordet–Gengou media
• direct pertussis antigen detection on nasopharyngeal aspirates.

Treatment

Most children under 1 year of age, and older children with complications, will need to be hospitalized.

• The older child without complications is allowed to be up and about. Support and reassurance are needed. Meals are kept small and frequent and, if vomiting is persistent, should be given after a paroxysm
• Erythromycin for 2 weeks eliminates B.pertussis. The newer macrolides are also effective, although experience is less.
For those admitted to hospital, the nursing management is vital.
• All cases should be isolated where possible
• Oxygen saturation should be monitored and oxygen should be given for recurrent or sustained hypoxaemia
• Gentle nasal suction should be performed to remove secretions
• Adequate hydration and nutrition need to be maintained, intravenously if necessary
• Disturbance of any sort should be kept to a minimum to reduce paroxysms
• Secondary bacterial pneumonia requires parenteral antibiotic treatment (e.g. cefuroxime)
• Major pulmonary collapse requires prolonged and intensive physiotherapy.
Steroids, β-agonists and sodium cromoglycate have no effect on the paroxysms.

Prevention

• Control of the disease is mainly by immunization. A full course of killed whole-cell vaccine (three doses at 2, 3 and 4 months) confers protection in over 80% of subjects for a decade. Less severe disease occurs in those not fully protected. Vaccines must contain all three major agglutinogens
• There is still no evidence of a causal relation between pertussis vaccine and permanent neurological illness, and confidence in the vaccine has improved (the vaccination rate has increased from 20% in 1980 to 87% presently in the UK). Nevertheless, acellular vaccines which have fewer side-effects have been developed and seem as efficacious. In the developing world only one-third of infants have been immunized
• Erythromycin given for 14 days to a contact will attenuate the illness if the child is in the catarrhal stage

• Erythromycin should be given to non- or partially immunized infant contacts.

Prognosis

Good nursing and medical practice reduce the complications, but morbidity is high. In England in 1982, more than 65 000 cases were reported, with 14 deaths (0.02%).

SARS (severe acute respiratory distress syndrome)

SARS

• Is an acute viral respiratory tract infection first identified in early 2003 in Hong Kong
• Is caused by a novel form of coronavirus (a family of viruses known to cause mild cold symptoms)
• Probably originated in southern China with rapid spread to Hanoi in Vietnam, Taiwan, Toronto and Singapore as a result of air travel. Isolated cases also occurred in other countries
• Has caused large numbers of cases in mainland China, Hong Kong and Singapore with secondary spread in the general population. In the West, Canada has been the worst affected and the only area where secondary spread has occurred
• Transmission:
   • is mainly via the respiratory route through transfer of infective respiratory secretion: droplet spread is more important than airborne spread and the majority of secondary cases are among close household or face-to-face contacts (e.g. health-care workers)
   • may possibly occur through faecal contact (e.g. via drains)
   • is less efficient than that of influenza, otherwise there would have been thousands more cases, with rapid global spread
• Clinical infection has not been reported since July 2003. Although a global pandemic seemed possible in early 2003, rapid and strict isolation measures brought the outbreak under control. It is possible an animal reservoir has been established and may trigger future outbreaks

- IP is around 7 days (range 2–10).

Clinical features
- Infection is characterized by sudden-onset high fever, myalgia, rigors and dry cough
- 3–4 days after onset of symptoms, characteristic patchy changes are visible on CXR with worsening of condition
- 80–90% show improvement on day 6–7
- A small number of patients go on to develop ARDS and require ventilatory support. Mortality associated with this type is high

Diagnosis and treatment
- Diagnosis is initially clinical, based on symptoms and exposure history, and exclusion of other causes, and supplemented by positive antibody/PCR tests against the new coronavirus strain. However, a negative test does not rule out SARS
- Available antivirals do not appear to be effective.

Prevention
- Patients must be isolated using full universal precautions with full protective gowns, gloves, goggles and masks, preferably in negative-pressure isolation cubicles, but the latter may become impractical in epidemic situations, particularly in resource-poor countries
- Strict quarantine of close contacts is not essential but they should be warned of the symptoms of SARS and advised of the importance of contacting the doctor immediately in the event of illness. In the event of epidemic spread into the general population, restrictions on people movement and congregation, and quarantine of large number of people or population groups may be necessary
- Useful websites for up-to-date information:
  - www.cdc.gov/ncidod/sars
  - www.who.int.en
  - www.doh.gov.uk/sars.

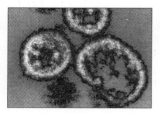

Infections of the Cardiovascular System

Clinical syndromes

Endocarditis, myocarditis and pericarditis

Major presenting features: *breathlessness, chest pain, fever.*

Causes

Endocarditis

Predisposing
• Rheumatic heart disease, congenital (e.g. bicuspid aortic valve, ventricular septal defect (VSD), patent ductus arteriosus (PDA), degenerative valvular disease, mitral valve prolapse, prosthetic valve)
• Intravenous (IV) drug abuse, preceding dental or other instrumentation.

Microbiological — native valve (80%)
• STREPTOCOCCUS VIRIDANS (*S. sanguis, S. mitior, S. bovis, S. milleri, S. mutans, S. salivarius*), groups C and G β-haemolytic streptococci, B_6-dependent streptococci
• STAPHYLOCOCCUS AUREUS, ENTEROCOCCUS FAECALIS, *Staphylococcus epidermidis*, 'coliforms'
• *Coxiella burnetii, Chlamydia, Legionella, Mycoplasma, HACEK bacteria*
• *Streptococcus pyogenes* (rheumatic fever).

Microbiological — prosthetic valve (20%)
• STAPHYLOCOCCUS EPIDERMIDIS, *S. aureus*, diphtheroids, *E. faecalis*, 'coliforms', *Candida.*

Microbiological — IV drug abuser
• STAPHYLOCOCCUS AUREUS, Pseudomonas aeruginosa, 'coliforms', *Candida, E. faecalis, Streptococcus viridans.*

Myocarditis
• COXSACKIE VIRUS, ECHOVIRUS
• Cytomegalovirus (CMV), mumps, Epstein–Barr virus (EBV), human immunodeficiency virus (HIV), influenza, adenovirus
• *Chlamydia, Coxiella, Mycoplasma pneumoniae*
• *Streptococcus pyogenes* (rheumatic fever), *Corynebacterium diphtheriae, Staphylococcus aureus* (ring abscess from endocarditis)
• Chagas' disease, trichinosis, toxoplasmosis, Lyme disease.

Pericarditis
• COXSACKIE VIRUS, ECHOVIRUS, mumps, EBV, influenza, adenovirus, HIV
• *Staphylococcus aureus, Streptococcus pneumoniae*, 'coliforms'
• Rheumatic fever (*Streptococcus pyogenes*)
• *Mycobacterium tuberculosis*, histoplasmosis, *Entamoeba histolytica.*

Distinguishing features
See Table 6.1.

Complications
Endocarditis: heart failure (acute/chronic), major embolus (carotid, coronary, limb), mycotic aneurysm, cardiac abscess, encephalopathy,

	Acute viral myocarditis	Bacterial endocarditis	Acute viral pericarditis
Incubation period	3–10 days	2–6 weeks	3–10 days
Season	Summer/autumn	All year	Summer/autumn
Chest pain	Occasional	Uncommon	Common
Dyspnoea	Prominent	Rare	Rare
Fever	Absent/low	Low grade	Absent/low
Tachycardia	Present	If acute	Uncommon
Heart murmur	Occasional	Present, new or changing	Absent
Gallop rhythm	Present	Absent	Absent
Signs of heart failure			
Left	Present	Uncommon	Absent
Right	Present	Uncommon	If effusion
Arrythmias	Common	Rare	Rare
Pericardial rub	Absent	Absent	Present
Immunological lesions*	Absent	Often present	Absent
Splenomegaly	Rare	50%	Rare
Clubbing	Absent	20%	Absent
Emboli	Absent	Occasional	Absent
Blood cultures	Negative	Positive	Negative
Microscopic haematuria	Absent	50%	Absent
Anaemia	Rare	Common	Rare
Chest X-ray			
Cardiomegaly	Often	Uncommon	If effusion
Left ventricular failure	Often	Uncommon	Absent
Electrocardiogram			
T waves	Inverted	Normal	Concave rise
QRS complex	May be widened	Normal	Normal
Echocardiogram	Poor left ventricular function	Vegetations in half	Normal or effusion

* Osler's nodes, splinter haemorrhages, Janeway lesions, Roth spots.

Table 6.1 Distinguishing features of infections of the heart.

glomerulonephritis, pericarditis, right–left fistulae.

Myocarditis: heart failure (acute), arrythmias, dilated cardiomyopathy.

Pericarditis: effusion, tamponade, constriction (tuberculosis).

Investigations

For all three infections
• Full blood count (FBCt), differential white cell count (WCCt) and erythrocyte sedimentation rate (ESR)
• Biochemical profile (liver function tests (LFTs), albumin and urea/creatinine)
• Chest X-ray (CXR)
• Cardiac enzymes to exclude myocardial infarction (MI)
• Electrocardiogram (ECG)
• Echocardiogram.

Endocarditis
• Blood cultures (at least three)
• Midstream urine (MSU):
  • blood and protein
  • microscopy and culture
• Serology:
  • *C. burnetii* (phase 2)

- full screen of viral and other atypical agents (if negative cultures)
- Minimum inhibitory/bactericidal concentrations (MICs/MBCs) (to confirm antibiotic sensitivities/choice)
- Serum back-titrations (to monitor therapy).

Myocarditis and pericarditis
- Viral cultures: throat swab and faeces
- Serology
  - antistreptolysin O (ASO) titre (rheumatic fever)
  - full screen of viral and other atypical agents
- Pericardial aspirate (microscopy, Gram and Ziehl–Neelsen (ZN) stains, culture and cytology)
- Biopsy:
  - endomyocardial to confirm and identify cause of myocarditis
  - pericardial to identify cause of pericarditis with effusion.

Differential diagnoses
Endocarditis: non-bacterial vegetations as seen in malignancy, systemic lupus erythematosus (SLE) (Liebman–Sacks), chronic wasting illnesses, uraemia; seronegative arthritides (e.g. ankylosing spondylitis, Reiter's syndrome); tertiary syphilis; atrial myxoma.
Myocarditis: sarcoidosis, cardiomyopathy, uraemia, radiation, drugs (doxorubicin, emetine), lead poisoning, toxaemia, MI, giant cell myocarditis.
Pericarditis: MI, Dressler's syndrome, uraemia, hypothyroidism, connective tissue diseases (Still's disease, SLE, rheumatoid disease), lymphoma/leukaemia, carcinoma, post surgery, trauma, post radiation.

Treatment
All three infections need bed rest, and antiarrythmic and antiheart failure drugs as indicated.

Endocarditis
- IV antibiotics:
  - *S. viridans*: benzylpenicillin
  - *S. aureus*: flucloxacillin
  - *E. faecalis*: ampicillin

- with initial gentamicin for 2 weeks (*S. viridans*) or 4 weeks (*E. faecalis* and *S. aureus*) followed by oral antibiotics for 4 weeks
- Surgery in the case of acute valvular incompetence, prosthetic valve disease or cardiac abscess.

Myocarditis
- Antimicrobial agent(s) if specific cause identified
- Immunosuppressives if severe and presumed of viral aetiology.

Pericarditis
- Antimicrobial agent(s) if specific cause identified (e.g. rifampicin, isoniazid, ethambutol, and pyrazinamide for tuberculous pericarditis)
- Aspirin or other non-steroidal anti-inflammatory drug (NSAID) (steroids are sometimes used) if presumed viral
- Aspiration if large effusion or tamponade present or imminent.

Rheumatic fever
- Aspirin (corticosteroids if severe).

Prevention
Endocarditis
- Antibiotic prophylaxis should be given during procedures (dental, surgical) for patients at risk (predisposing causes above) (see p. 68).

Myocarditis/pericarditis with rheumatic fever
- Secondary prophylaxis with oral penicillin V or monthly benzathine penicillin G
- Primary prophylaxis by treating *S. pyogenes* tonsillopharyngitis with penicillin V.

Specific infections

Endocarditis
Epidemiology
Infective endocarditis results from an infection of the cardiac valves. Although usually subacute and due to a bacterium, it may be acute and have a non-bacterial aetiology, hence the preference

for the term 'infective endocarditis' rather than 'subacute bacterial endocarditis'. Infective endocarditis:

• has an incidence of 6–7/100 000, resulting in 3000–4000 cases/year in the UK
• usually (60–80%) involves a previously abnormal value — rheumatic heart disease (25%), congenital abnormalities (e.g. bicuspid aortic valve, VSD), degenerative disease (aortic and mitral valve sclerosis), mitral valve prolapse and prosthetic valves
• is increasing in frequency in developed countries. This is the result of an increasing elderly population with degenerative valvular disease, more invasive intravascular procedures, longer survival of children with congenital heart disease, increasing IV drug abuse and increasing numbers of patients with prosthetic valves. The disease is rare in children
• occurs mainly in the elderly (50% > 50 years of age)
• most frequently affects the mitral and aortic valves, either alone (45% and 30%, respectively) or together (20%). In IV drug use (IVDU), the commonest valve affected is the triscuspid (50%)
• may affect previously normal valves, as is the case in 80% of injection drug users
• affects patients with a history of antecedent dental treatment in 15% of cases
• has an IP of 2–6 weeks.

Pathology and pathogenesis

Classically, platelets and fibrin initiate a vegetation on endothelial breaks which are then colonized by a circulating organism. The bacteraemia follows on from trauma to a colonized mucosal surface, usually the mouth. The vegetation develops at sites lateral to maximum turbulence (small defects produce more turbulence than large ones) and downstream to the defect (e.g. atrial side if atrioventricular valve and ventricular side if semilunar valve). Hence, patients with mitral stenosis or an atrial septal defect (ASD) rarely develop endocarditis, whereas those with aortic incompetence or a VSD are prone to. Where endocarditis develops on previously normal valves, the organism tends to be more

virulent: destruction of the valve is rapid and may lead to perforation of the cusp or rupture of the chorda tendinae or papillary muscle. The clinical features result from embolism and/or immune complex deposition. Major vessel infarcts, most mycotic aneurysms and microscopic haematuria, arise from embolization with infarction. Vasculitic skin lesions, splinter haemorrhages, Roth spots (retina), Osler's nodes and glomerulonephritis are predominantly immunological. Circulating immune complexes are seen only where infection has been subacute.

Clinical features

The classic features are:
• fever (usually low grade), drenching sweats, influenza-like symptoms (myalgia, arthralgia, malaise) and weight loss
• clubbing (20%), splenomegaly (50%) and anaemia (80%)
• changing heart murmur. A murmur may be absent in a small minority (especially right-sided endocarditis in IVDU)
• splinter haemorrhages (60%), vasculitic skin lesions (Janeway spots) (5%), Osler's nodes (10%), Roth spots (5%), mucous membrane petechiae (30% — conjunctivae and palate)
• evidence of major embolism. In right-sided disease, the infarcts are pulmonary.

In acute endocarditis, hectic fever, rigors and toxaemia are frequent, whereas skin features are rare. In early-onset prosthetic valve endocarditis, the course is often rapid with rapid valvular dysfunction: late onset is usually indistinguishable from native valve subacute disease.

Endocarditis should always be considered as a possible diagnosis in many presentations, particularly in patients:
• with a pyrexia of unknown origin (PUO) and heart murmur
• with unexplained cardiac failure, embolic episodes or a 'vasculitic' rash
• where S. viridans has been isolated from a blood culture
• with a prosthetic valve
• who are IV drug abusers.

Complications

Cardiac

- Cardiac abscess
- Cusp perforation, ruptured chorda tendinae or papillary muscle
- Coronary artery embolus
- Acute or subacute ventricular failure
- Pericarditis.

Systemic

- Embolism (coronary, carotid, splenic, renal or limb)
- Mycotic aneurysm
- Glomerulonephritis (focal, diffuse and membranoproliferative)
- Encephalopathy.

Diagnosis

Non-specific features supporting a diagnosis of endocarditis include:

- low-grade fever, heart murmur, 'vasculitic' lesions, splenomegaly
- normocytic normochromic anaemia, elevated ESR, microscopic haematuria
- multiple infarcts on CXR (right-sided disease).

Confirmation of endocarditis is by:

- demonstration of vegetations by echocardiography (seen in 50%). The sensitivity can be increased by using a transoesophageal probe (90%): it is particularly useful in the diagnosis of prosthetic valve endocarditis and myocardial abscess
- recovery of an organism from blood cultures:
  (a) in native valve endocarditis:
  - S. viridans (40%), S. aureus (25%) and E. faecalis (20%) in those with subacute disease (who are not IV drug abusers); S. aureus in those with acute endocarditis
  - S. aureus, P. aeruginosa, Candida (large vegetations) and 'coliforms', or the isolation of more than one organism from IV drug abusers
  - no organisms are isolated in 10% (culture-negative cases). This may be due to recent antibiotic administration or the presence of pyridoxine-dependent S. viridans, C. burnetii, fungi or other fastidious microorganisms

such as the HACEK group of bacteria (Haemophilus, Actinobacillus, Cardiobacterium, Eikenella and Kingella)
(b) in prosthetic valve endocarditis:

- disease is divided into 'early' (within 60 days) and 'late' infection
- 'early' endocarditis indicates infection acquired at the time of surgery; the causes are S. epidermidis (30%), S. aureus (20%), Gram-negative bacilli (18%), diphtheroids (10%) and Candida (10%)
- 'late' endocarditis has a similar microbiological aetiology to native valve disease.

Treatment

Endocarditis is potentially curable: this is more likely the earlier treatment is initiated.

- IV antibiotics:
  - S. viridans: benzylpenicillin and gentamicin (2 weeks)
  - S. aureus: flucloxacillin and gentamicin (4 weeks minimum)
  - E. faecalis: ampicillin and gentamicin (4 weeks minimum)
  - S. epidermidis (prosthetic valve): vancomycin and additional antibiotics depending on sensitivity pattern (6 weeks minimum)
- Oral antibiotics (4 weeks minimum) after initial IV treatment:
  - S. viridans: amoxicillin
  - S. aureus: flucloxacillin
  - E. faecalis: amoxicillin
- Determining the MIC/MBC of the organism is essential in determining antibiotic regimen and may influence the duration of initial IV treatment
- Back-titrations assist in confirming effective bactericidal drug concentrations in serum
- Surgery may be needed for acute valvular incompetence, prosthetic valve disease (to replace prosthesis), cardiac abscess, or when cultures remain positive or become positive again despite optimal antibiotic therapy
- Bed rest, antiarrythmic and antiheart failure drugs as indicated
- Recurrence of fever may indicate myocardial abscess, drug hypersensitivity, secondary infection, or an alternative diagnosis.

Prevention

- At-risk patients undergoing procedures liable to be complicated by bacteraemia should be given antibiotic prophylaxis. Regimens depend upon the type and site of procedure and whether or not the patient has had a previous attack of endocarditis; has a prosthetic valve; or has received penicillin during the previous month. Recommendations are based on the following antibiotics: amoxicillin (or erythromycin or clindamycin if penicillin-allergic); amoxicillin and gentamicin (genitourinary procedure and other complicated cases); and vancomycin and gentamicin (complicated cases where penicillin-allergic or recently received penicillin)
- Patient education.

Prognosis

In typical subacute endocarditis due to *S. viridans*, mortality is only 5%. In acute endocarditis due to *S. aureus* or prosthetic valve-associated endocarditis, mortality is much higher (early 50%, late 25%). Factors associated with poor prognosis include non-streptococcal disease, heart failure, aortic valve involvement, prosthetic valve disease, older age and myocardial abscess. About 10% of patients have second attacks later in their life.

Rheumatic fever

Epidemiology

Rheumatic fever is a multisystem disease which follows on from an *S. pyogenes* throat infection in the preceding 2–4 weeks. It is characterized by migratory arthritis and pancarditis.

- It results from tonsillopharyngeal infection with rheumatogenic M-types of *S. pyogenes* (e.g. M-type 18); it is not seen complicating infection at other sites. Incidence is closely linked to that of *S. pyogenes* tonsillitis
- It has an incidence of up to 1/1000 in developing countries, 100 times more common than in the West. It is a disease of overcrowding and low socioeconomic status
- It occurs in 3% of patients with untreated streptococcal tonsillitis (if correct serotype)
- It is a disease of childhood (5–15 years) when it may be missed, patients later presenting with

rheumatic heart disease. Infection does occur in young adults but it is rare
- Case numbers have been increasing recently in Western countries associated with a return of rheumatogenic M-types (e.g. 5, 6, 18—see p. 32)
- It is more severe, and progresses more rapidly, in developing nations reflecting earlier age of first infection and more recurrences
- It has an IP of 2–4 weeks.

Pathology and pathogenesis

The pathognomic lesion of rheumatic fever is Aschoff's nodule, a perivascular aggregate of lymphocytes and plasma cells surrounding a fibrinoid core. The basic disease process is that of an inflammatory vasculitis. In the heart, mitral valvulitis is the most common lesion, associated with small, firmly adherent vegetations. The acute pathological changes are followed by fibrosis and subsequent deformity of the valve cusps and chorda tendinae, leading to valve stenosis and/or incompetence. Rheumatic fever results from an aberrantly hyperreactive immune response to *S. pyogenes* antigens. Both cell-mediated and humoral responses are exaggerated and a genetic susceptibility to the disease is likely.

Clinical features

Because the diagnosis is based on clinical findings, these have been well characterized.

- Arthritis is the most common manifestation (80% of cases) and is migratory, large joint, asymmetrical and non-deforming. Pain and tenderness exceed redness and swelling. As one joint improves, another becomes affected; the whole process lasts 3–6 weeks. The response to aspirin is dramatic
- Carditis occurs in half of cases, and may be asymptomatic. It may lead to subsequent rheumatic heart disease. Mitral and aortic valvulitis, myocarditis and pericarditis may all occur
- Chorea is a late manifestation (2–6 months) and can occur alone or with carditis (which may persist for 6 months), affecting 10% overall. It is more common in girls aged 8–12 years

• Erythema marginatum (occurring in 5%) is a non-pruritic skin rash, appearing as pink macules which clear centrally leaving a rim of red which spreads
• Subcutaneous nodules are round, firm and painless and appear over bony prominences such as the occiput and elbow. They are rare in first attacks
• Low-grade fever is usually present.

Complications
• Pancarditis
• Recurrent rheumatic fever
• Rheumatic heart disease.
Rheumatic heart disease is the major complication. The probability of recurrent rheumatic fever after further S. pyogenes infection is 65%. Approximately one-third of patients who have rheumatic fever (one-half of those with carditis) develop rheumatic heart disease, and severe disease results from recurrent attacks. The mitral valve is affected in nearly all patients, the aortic valve in 30% and the triscuspid in 15% (always accompanied by mitral valve disease).

Diagnosis
The diagnosis is based upon fulfilling certain clinical criteria and showing recent streptococcal infection. The modified Duckett Jones criteria are split into 'major' and 'minor'. Either two major, or one major and two minor, criteria are required.
• The five 'major' criteria are carditis, arthritis, chorea, erythema marginatum and subcutaneous nodules
• The five 'minor' criteria are fever, arthralgia, previous rheumatic fever or heart disease, raised acute phase reactants (ESR, C-reactive protein, leucocytosis) and prolonged PR interval
• Evidence for recent S. pyogenes infection needs to be provided by a positive throat culture, increased ASO titre or other anti-streptococcal antibodies (e.g. anti-DNAse or antihyaluronidase), or a history of recent scarlet fever.

Treatment
• Bed rest if carditis present
• Aspirin or, in severe disease, corticosteroids
• Penicillin to eradicate pharyngeal S. pyogenes.

Prevention
• Secondary prophylaxis with oral penicillin V or monthly benzathine penicillin G. This is essential in preventing recurrences
• Primary prophylaxis by treating S. pyogenes tonsillopharyngitis with penicillin V
• There is no vaccine.

Prognosis
Mortality in the initial attack is very rare (< 1%) and results from severe carditis. One-third progress to develop rheumatic heart disease which has a significant chronic morbidity and mortality.

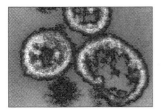

Infections of the Nervous System

Clinical syndromes

Acute meningitis

Major presenting features: *rapidly developing headache, fever, meningism and photophobia.*

The most important and immediate task of the clinician is to distinguish between pyogenic bacterial, viral and tuberculous meningitis.

Causes

Acute bacterial (pyogenic)
• Neonates: ESCHERICHIA COLI, GROUP B STREPTOCOCCI, *Proteus mirabilis, Pseudomonas* (if hospital acquired), *Listeria monocytogenes*
• Children <5 years: NEISSERIA MENINGITIDIS, *Streptococcus pneumoniae, H. influenzae* b (commonest cause in countries without Hib vaccination programme)
• Older children and adults <50 years: N. MENINGITIDIS, *S. pneumoniae*
• Adults >50 years: S. PNEUMONIAE, Gram-negative enteric bacilli, *L. monocytogenes.*

Viral (aseptic)
• ENTEROVIRUSES, MUMPS (rare in developed countries due to routine childhood vaccination), arboviruses (in some countries)
• Human immunodeficiency virus (HIV), herpes simplex virus 2 (HSV-2) (uncommon), herpes zoster, adenoviruses.

Tuberculous
• *Mycobacterium tuberculosis.*

Distinguishing features
See Table 7.1.

Complications

Pyogenic: deafness, blindness, cranial nerve palsies, hydrocephalus, subdural haematoma, cerebral abscess, Waterhouse–Friederichsen syndrome (meningococci), intellectual impairment.
Viral: none.
Tuberculous: hydrocephalus, blindness, hemiparesis.

Investigations
• Gram stain of cerebrospinal fluid (CSF) smear
• CSF smear stain for acid-fast bacilli (AFB) (if tuberculosis is suspected), biochemistry

SAH

	Pyogenic[1]	Viral[2]	Tuberculous[3]
Onset	Acute (< 2 days)	Acute (< 2 days)	Subacute (> 2 days)
Clinical features	Toxic and ill, drowsy, possible purpuric rash (meningococci)	Not toxic, fully conscious[4]	Not toxic, alertness may be depressed
Cerebrospinal fluid (CSF)			
Appearance	Turbid or opalescent	Clear	Clear, may form cobweb on standing
Cells (number/μL)	Polymorphs (500–2000)	Lymphocytes (5–1000)[5]	Lymphocytes (50–400)[6]
Protein	Increased	Normal or modest rise	Increased
Glucose	Reduced	Normal[7]	Reduced
Gram stain	Usually positive	Negative	Negative
White cell count	Neutrophilia	Normal	Normal

1 A brain abscess rupturing into subarachnoid space will produce a similar CSF picture.
2 A similar CSF picture can be due to encephalitis, intracranial sepsis, Lyme disease, leptospirosis, syphilis, early tuberculosis and non-infective causes (e.g. sarcoidosis, cerebral lupus).
3 Cryptococcal and other fungal meningitis will cause similar CSF changes; some cases of listeria meningitis may present subacutely and have lymphocytic CSF and reduced glucose.
4 Parotitis is often present in cases due to mumps virus. A maculopapular rash may accompany enterovirus meningitis.
5 CSF may initially be more neutrophilic in enteroviral infection.
6 CSF may be mixed lymphocytic/neutrophilic.
7 Glucose may be modestly reduced in mumps meningitis.

Table 7.1 Differential diagnosis of acute meningitis.

• Blood culture and white cell count (WCCt), plasma viscocity
• Chest X-ray (CXR)
• Demonstration of antigen in CSF/blood (by countercurrent immunoelectrophoresis (CIE) or Latex agglutination) if Gram stain is negative in pyogenic meningitis
• Faecal, throat swab and CSF viral cultures and viral serology in all cases of aseptic meningitis.

Diagnosis

Bacterial
• Typical CSF picture
• Identification of organism from CSF Gram stain, CSF culture, antigen detection in CSF/blood.

Viral
• Enterovirus: identification in faeces, CSF, throat swab
• Mumps: isolation from CSF, urine; serology
• Arbovirus: serology, PCR of CSF.

Tuberculous
• AFB in CSF smear (in a minority)
• CSF polymerase chain reaction (PCR)
• CSF culture.

Investigation and treatment
Acute bacterial meningitis is a life-threatening emergency
Early diagnosis and treatment are the key to a successful outcome
• Lumbar puncture (LP) should be performed in all cases unless papilloedema or focal neurological signs are present
• If these signs are present, begin empirical an-

tibiotic therapy (see Table 3.1) after taking blood culture
- Empirical antibiotic therapy should also begin if:
  - patient looks ill or toxic, or
  - has a petechial rash, or
  - is at home and meningococcal infection is likely (give an injection of intravenous (IV) penicillin)
- LP yields turbid CSF: patient has bacterial meningitis; initiate empirical antibiotic therapy (see Chapter 3)
  - CSF Gram-stain reveals causative bacteria: review therapy (see individual infections)
  - Gram-stain negative: continue empirical therapy
- CSF is clear: await laboratory results
- CSF is lymphocytic, normal biochemistry: patient has probable viral meninigitis; review likelihood of other causes of similar CSF changes and arrange diagnostic virology
- CSF is lymphocytic, protein raised, glucose reduced, AFB or fungal stain positive: begin appropriate therapy
- CSF is lymphocytic, protein raised, glucose reduced, search for AFB/fungi negative: tuberculosis (TB) still likely; begin anti-TB therapy, cover for Listeria.

Acute encephalitis and brain abscess

Major presenting features: *evidence of cerebral dysfunction of acute onset.*

Causes

Acute encephalitis
- Primary: HERPES SIMPLEX, ENTEROVIRUS, ARBOVIRUSES, rabies, HIV, mycoplasma, Epstein–Barr virus (EBV), varicella zoster virus (VZV), cytomegalovirus (CMV), influenza
- Post-exanthemata: measles, rubella, varicella, vaccinia, acute disseminated encephalomyelitis (ADEM).

Brain abscess
- STREPTOCOCCUS MILLERI, 'COLIFORMS', ANAEROBES, other streptococci
- *P. aeruginosa* (otitic source), *S. aureus* (blood-borne, trauma, neurosurgery)

- *S. pneumoniae* (complicating meningitis)
- *Toxoplasma gondii* (in immunocompromised).

Distinguishing features
See Table 7.2.

Diagnosis
- The diffuse cerebral dysfunction, CSF abnormalities and absence of focal neurological signs will suggest the diagnosis of encephalitis and search for viruses should be initiated
- Computed tomography (CT) or magnetic resonance (MR) scans will be diagnostic of brain abscess or corroborate the diagnosis of encephalitis, may point towards HSV encephalitis or ADEM, or may reveal a non-infective cause.

Differential diagnoses
- Toxic confusion in systemic infection
- Cerebral malaria
- Subdural haematoma
- Tuberculous meningitis
- Neurocysticercosis
- Subarachnoid haemorrhage
- Cerebrovascular accidents.

Treatment
Encephalitis
- IV acyclovir if HSV is possible, otherwise nutritional and cardiorespiratory support
- Treatment of convulsion; dexamethasone to relieve raised intracranial pressure (ICP).

Brain abscess
- Empirical antibiotic therapy with metronidazole, ampicillin and cefotaxime (ceftazidime if otitic)
- Surgery.

Acute polyneuritis, acute myelitis and polio-like illness

These conditions may present in a similar manner. Major presenting features: *rapidly developing ascending weakness of limbs and trunk.*

Causes
Acute polyneuritis
- GUILLAIN–BARRÉ SYNDROME, diphtheria, postvaccinial.

	Encephalitis	Brain abscess
Presentation	Acute	Subacute
Fever	Common and prominent	Common (low grade)
Headache	Common	Common
Nausea and vomiting	Common	Common
Meningism	Common in enteroviral and arboviral infections	Rare
Papilloedema	Rare	Common
Mental status	Commonly confused, delirious and disorientated	Only apathy and inattention in early stages Deteriorating mental function later
Seizures and focal signs	Unusual except in HSV encephalitis	Common
Laboratory:		
White cell count	Unremarkable	Neutrophila (often low grade) common
Cerebrospinal fluid	Similar to viral meningitis	Normal, or mild mixed cellularity (depends on nearness to sub arachnoid space), normal biochemistry or mildly raised protein NB: lumbar puncture should not be performed if brain abscess is suspected
CT/MR scan of head	Oedematous brain ± mass effect	Hypodense shadow with ring enhancement and surrounding oedema
	Contrast-enhancing areas in temporal lobes may be present in HSV infection High signal intensity affecting white matter extensively in ADEM (MR scan)	Daughter abscesses
Antecedent events	Usually none in primary encephalitis	Source of infection may be evident (ear or sinus infection, pneumonia, bacteraemia
	Recent exanthema or nerve tissue vaccination in ADEM	Skull fracture, neurosurgery AIDS

Table 7.2 Distinguishing features of acute encephalitis and brain abscess.

Acute myelitis
- IDIOPATHIC, HSV-2, herpes zoster, CMV, EBV
- Lyme borreliosis, schistosomiasis.

Polio-like illness
- Poliovirus, coxsackievirus and echovirus
- Vaccine-derived poliovirus
- West Nile virus.

Distinguishing features
See Table 7.3.

Investigations
Gullain–Barré syndrome
- Nerve conduction studies (if diagnosis in doubt)
- Exclude diphtheria (history of immunization, nasopharyngeal swab)
- Investigate for *Campylobacter*, EBV and CMV infection.

	Guillain–Barré syndrome	Acute myelitis	Polio-like illness
Neurological findings			
Muscle weakness			
Nature	LMN	UMN	LMN
Distribution	Ascending, symmetrical	Below a level in trunk	Asymmetrical
Sensation	Diminished	Absent below a level in trunk	Absent
Reflexes	Absent	Exaggerated	Absent in severe weakness
*Cerebrospinal fluid**			
Cells/µL	<5	5–200, mononuclear	5–500
Protein	Very high	Slightly raised	Normal or slightly raised
Glucose	Normal	Normal	Normal

* Lumbar puncture should not be performed in suspected myelitis without first excluding spinal cord compression (which produces similar weakness below a level in trunk) by magnetic resonance scan, as lowering of CSF pressure may aggravate paralysis. LMN, lower motor neurone; UMN, upper motor neurone.

Table 7.3 Differential diagnosis of acute muscular weakness of infective origin.

Acute myelitis
• Investigate for HSV, VZV, CMV, Lyme borreliosis
• Consider schistosomiasis if relevant travel history
• Spinal X-ray and MR scan.

Polio-like illness
• Virus isolation from faeces, throat swab, CSF, serology, PCR
• If poliovirus isolated, confirm whether wild or vaccine derived.

Differential diagnosis
Botulism, paralytic shellfish poisoning and acute spinal cord compressions (e.g. from spinal epidural abscess, spinal tuberculosis, malignancy) are other important causes of acute-onset paralysis of limbs and trunk.

Treatment
• Polio-like illness: bed rest, respiratory support, physiotherapy; orthopaedic appliances and later corrective surgery

• Myelitis: bed rest, rehabilitation
• Guillain–Barré syndrome: consider plasmapheresis or IV immunoglobulin.

Specific infections

Meningococcal meningitis and septicaemia
Epidemiology
• Meningococcal meningitis is the commonest form of bacterial meningitis everywhere with at least 500 000 cases annually worldwide with 50 000 deaths
• It is the only form of bacterial meningitis that causes epidemics
• It is caused by *Neisseria meningitidis*, a Gram-negative intracellular diplococcus; recognizable pathogenic groups are A, B, C, D, X, Y, Z and W135
• The disease is endemic in temperate climates, causing a steady number of sporadic cases due to groups B and C, with a seasonal increase in winter and spring
• Group C also causes occasional small outbreaks in households, nurseries, schools and other educational settings

- Around 2500 cases occur annually in the UK, about two-thirds due to group B, one-third due to group C and <5% due to other groups
- The incidence of group C cases has declined dramatically in the UK since the recent introduction of group C vaccination for infants and 15–17-year-olds
- The highest burden of meningococcal disease is in sub-Saharan Africa (from Ethiopa to Senegal—'meningitis belt') where it is both endemic and epidemic, particularly in the dry season and involves groups A, C and W135. In Asia group A is usually involved
- In temperate climates, the majority of infections occur in infants <5 years (highest incidence in infants < 1 year) with a secondary peak in 15–19-year-olds
- The organism is carried asymptomatically in the nasopharynx in approximately 10% of the population, with the highest carriage (up to 25%) in 15–19-year-olds
- Meningococci are transmitted by droplet spread or direct contact from carriers or from patients in the early stages of infection. Close prolonged contact in a household setting is usually necessary. Overcrowding encourages transmission.

Pathogenesis

- Initially there is colonization of nasopharyngeal mucosa. Most cases do not progress further
- In others, there is local invasion followed by bacteraemia and rapid intravascular multiplication
- Meningeal invasion leads to subarachnoid space inflammation, or
- The patient may present purely as septicaemia with rapidly progressive shock and disseminated intravascular coagulation with bleeding into and dysfunction of many organs of the body, including the adrenals (Waterhouse–Friederichsen syndrome)
- The characteristic purpuric rash results from as yet poorly understood local factors triggered by meningococci, compounded by the rapidly developing thrombocytopenia
- Group-related immunity follows an infection,

even if subclinical, but nasopharyngeal carriage may continue
- Incubation period (IP) is 1–3 days.

Clinical features

- Illness starts abruptly with fever, vomiting, headache, irritability and restlessness, rapidly progressing to signs of meningitis, or
- May present just as fulminant septicaemia with rapidly developing toxicity, drowsiness and shock
- A petechial or purpuric rash is present in two-thirds of all cases. Initially it may be macular, becoming rapidly petechial and extensive
- Babies may just look acutely ill with a blotchy or pale skin, refusing to feed, fretful and crying and may have a stiff, jerking body
- Rarely, a patient may present with chronic bacteraemia with recurrent bouts of fever and rash but otherwise be relatively well.

Complications

- Waterhouse–Friderichsen syndrome: fulminant septicaemia with evidence of adrenocortical failure. Tissue damage resulting from ischaemia may be severe with loss of fingers or toes
- Hydrocephalus, brain damage, subdural haematoma, brain abscess, deafness
- Arthritis (septic or reactive), cutaneous vasculitis, pericarditis.

Diagnosis

- A high index of suspicion is essential whenever dealing with a patient with meningism or a petechial rash
- Confirmation is by CSF examination or blood culture
- CSF changes are usually characteristic of pyogenic meningitis (see Table 7.1) and Gram stain shows Gram-negative diplococci, often intracellular
- Gram stain may be negative with prior antibiotic therapy, so CSF and blood cultures are essential. Antigen detection in CSF and blood (Latex agglutination, CIE) are often helpful for rapid diagnosis in Gram-stain-negative cases

- PCR helps detection of meningococci in blood and CSF.

Treatment
- Petechial rash present: IV penicillin 10 days
- Petechial rash absent: IV cefotaxime until diagnosis confirmed, then change to IV penicillin or continue
- Septicaemia: ill patients should have haemodynamic monitoring, with cardiorespiratory support if necessary
- Patient should receive rifampicin or ciprofloxacin before discharge to eradicate nasopharyngeal carriage (unaffected by penicillin or cefotaxime).

Prognosis
The overall mortality is 5–10%, mostly from rapidly progressive septicaemia.

Prevention
- All cases must be notified and chemoprophylaxis given to all close contacts (household contacts, contacts in daycare centres): either rifampicin 600 mg twice daily for 2 days (children 10 mg/kg) or ciprofloxacin (500 mg single dose in adults) or single injection of ceftriaxone
- Nasopharyngeal swabbing is not necessary for prevention purposes
- Chemoprophylaxis will not prevent illness in those already incubating it (coprimary cases) and close surveillance of intimate contacts is important
- Meningococcal group C conjugate vaccine gives long-term protection against that group and is now routinely given as part of the childhood immunization schedule, also to students prior to entering higher education institutions and to asplenic individuals
- A meningococcal polysaccharide vaccine against groups A, C, W135 and Y is now available and is recommended for travellers going to high incidence areas. This is poorly immunogenic in children < 2 years
- Vaccine is currently unavailable against group B infections.

Haemophilus meningitis
Epidemiology and pathogenesis
- *Haemophilus influenzae* organisms are common in the respiratory tract of both children and adults but, of the six antigenically distinct capsular types (a, b, c, d, e and f), only type b produces meningitis
- Possible inflammation of the upper respiratory passages occurs initially, followed by invasion of the bloodstream, with involvement of meninges hours or days later
- Transmission is via respiratory route. Cases mainly occur in children from 3 months to 5 years of age with a peak incidence around 1 year of age. Older children and adults are immune because of previous, often subclinical, infections
- Cases are usually sporadic but household spread may occur with secondary cases among susceptible children
- The incidence of invasive *H. influenzae* disease has declined dramatically in countries that have introduced conjugated *H. influenzae* type b (Hib) vaccine. It is now rare in the UK.

Clinical features
- The disease often presents less acutely than either meningococcal or pneumococcal meningitis
- A period of fever and malaise often precedes the development of signs of meningeal irritation. Untreated, the patient will become drowsy and eventually comatose, sometimes with convulsions and cranial nerve paralysis.

Diagnosis
- Lumbar puncture usually reveals a CSF typical of pyogenic meningitis (see Table 7.1). The Gram stain shows Gram-negative pleomorphic coccobacilli and CSF and/or blood cultures are usually positive. The bacterial antigen can usually be detected by CIE in Gram-negative cases.

Complications
- Cranial nerve palsies
- Hydrocephalus

• Deafness, particularly in cases of delayed treatment
• Subdural effusion.

Treatment
• Ampicillin- and chloramphenicol-resistant strains are not uncommon and a third-generation cephalosporin such as cefotaxime, cefuroxime or ceftriaxone is the drug of choice
• Dexamethasone as an adjunctive therapy helps to prevent deafness.

Prognosis
Effective early treatment reduces the mortality to almost 1%. Complete recovery is usual but long-term neurological sequelae may occur if treatment is delayed.

Other diseases produced by H. influenzae type b
These also occur only in the young:
• Acute epiglottitis
• Septic arthritis/osteomyelitis
• Cellulitis affecting the face, head or neck. The onset is abrupt and the affected skin often assumes a bluish-purple appearance, which may be mistaken for a bruise
• Pneumonia, empyema and pericarditis.

Prevention
• Consider prophylaxis (rifampicin 20 mg/kg daily for 4 days) if there are other children in the household below the age of 4 years. Similar action should be taken in daycare centres if two or more cases occur
• A protein–polysaccharide conjugate vaccine gives effective protection against H. influenzae type b diseases and is a part of the childhood immunization schedule in most developed countries.

Pneumococcal meningitis
Epidemiology and pathogenesis
• It is the second most common cause of bacterial meningitis in adults worldwide
• It is the commonest form of bacterial meningitis in patients >50 years, though not uncommon in young infants

• Asplenic and hypogammaglobinaemic patients are at higher risk
• The source of infection may be haematogenous from the respiratory tract or an unknown source, or direct from a chronically infected middle ear or via a congenital defect or skull fracture (may cause recurring disease)
• The disease is not infectious.

Clinical features
• Fever, headache, neck stiffness, photophobia and vomiting start acutely, progressing rapidly to drowsiness and coma, and sometimes convulsions
• In patients with recurrent pneumococcal meningitis, there may be a history suggestive of CSF rhinorrhoea or otorrhoea.

Complications, diagnosis and prognosis
• Cranial nerve damage, ventriculitis and hydrocephalus (commoner in pneumococcal meningitis than in other types of bacterial meningitis)
• Cerebral abscess
• CSF shows typical changes of pyogenic meningitis (see Table 7.1). Gram stain of CSF deposit usually shows Gram-positive diplococci.
• CIE for pneumococcal antigen is usually positive in CSF/serum and sometimes in urine. CSF and blood should be cultured for isolation and determination of antibiotic susceptibility
• Mortality is high (up to 20%) despite treatment.

Treatment and prevention
• Cefotaxime is the drug of choice. Multidrug resistance is spreading, so antibiotic suceptibility should be checked
• Patients with asplenia, diabetes, immunosuppression or heart, lung or kidney diseases and patients >65 years are at increased risk and should receive pneumococcal vaccine.

Group B streptococcal meningitis and septicaemia
Epidemiology
• Lancefield group B streptococcus (GBS) is a

leading infectious cause of neonatal morbidity and mortality
• GBS causes invasive diseases in the newborn, manifesting as septicaemia, meningitis or pneumonia
• In pregnant women, GBS can also cause septic abortion and puerperal sepsis, and rarely urinary infection, but is usually asymptomatic; urinary or wound infection may occur in the elderly
• Incidence of neonatal invasive GBS disease has declined in recent years in the UK, USA and other countries following widescale use of intrapertum antibiotic prophylaxis to women at risk of transmitting GBS to their neonates. It is now between 0.5 and 0.8/1000 live births
• The gastrointestinal (GI) tract is the natural reservoir for GBS and often leads to vaginal colonization (10–30% of pregnant women)
• Source of infection is either from mother's genital flora causing the 'early-onset disease' manifesting within 7 days of birth; or cross-infection in neonatal unit, usually causing the 'late-onset disease' which occurs between the 7th and 90th day of life
• Predisposing factors are preterm delivery, prolonged rupture of membranes in labour, chorioamnionitis.

Clinical features

Early-onset disease usually presents non-specifically with signs of sepsis or pneumonia; less frequently meningitis. Pallor, irritability, vomiting and failure to feed and thrive may be the only signs; jaundice and respiratory distress may be present. A bulging fontanelle is present in a minority of cases. All are septicaemic.

Late-onset disease presents more often with meningitis.

Diagnosis

• *Escherichia coli* is the other important cause of neonatal meningitis, less commonly *Listeria monocytogenes* and other Gram-negative bacilli
• CSF examination is of less value in neonatal meningitis than in older children, as cell counts up to 30/μL and protein levels up to 1.5 g/L are

not uncommon in high-risk infants without meningitis
• Gram stain of CSF deposit and culture of CSF and blood are more helpful.

Treatment

Empirical broad-spectrum antibiotic therapy should begin urgently on suspicion of neonatal meningitis or septicaemia, once blood and CSF samples have been collected. The combination of ampicillin with cefotaxime or an aminoglycoside is widely used.

Prevention

• Intrapartum antibiotic prophylaxis reduces the risk of early-onset disease in the newborn
• CDC now recommends universal perinatal screening for vaginal and rectal colonization of all pregnant women at 35–37 weeks' gestation, with intrapartum penicillin or ampicillin prophylaxis for those found positive, or who have had a previous infant with invasive GBS disease
• Those with premature rupture of amniotic membranes, or chorioamnionitis or intrapartum fever should receive broad-spectrum antibiotic therapy that includes GBS cover.

Prognosis

Modern neonatal care has reduced the previously high mortality of up to 50% to around 4%.

Listeriosis and *Listeria* meningitis
Epidemiology

• *Listeria monocytogenes* is a Gram-positive bacillus present in the environment (water and earth) which commonly infects humans and animals, leading to asymptomatic faecal excretion
• Vaginal carriage may occur in women
• Sources of human infection are unpasteurized milk, cheese, pâté, faecally contaminated vegetables or infected materials (e.g. aborted animal fetus)
• Neonatal listeriosis can be acquired from an infected maternal birth canal
• Neonates, pregnant women and immuno-compromised or otherwise debilitated persons

and individuals >50 years have a higher incidence
• The IP varies from 3 to 7 days.

Clinical features
• Infection is often asymptomatic in the normal host or may only produce a short-lasting influenza-like illness
• In pregnant women, transient bacteraemia of mild illness can still cause fetal infection leading to abortion or a stillborn or badly damaged baby
• Occasionally meningitis (mainly in adults) or septicaemia may develop
• Meningitis:
  • 20% of adult cases are neither pregnant nor immunocompromised
  • onset is typical of pyogenic meningitis but in the immunocompromised or elderly, the disease may present subacutely
  • CSF usually shows changes of pyogenic meningitis but may be lymphocytic and mimic tuberculous meningitis because of a low glucose level
  • neonates may have a septicaemic illness at birth, or meningitis later in the neonatal period.

Diagnosis
• Is by isolation of the organism by culture from CSF, blood, meconium and gastric washings. The organisms may be confused with diphtheroids because of similar morphology
• CSF Gram stain is usually negative.

Treatment and prognosis
• The drug of choice is ampicillin, combined with an aminoglycoside
• Co-trimoxazole is a suitable alternative for penicillin-allergic patients
• Mortality is high (30% or more).

Prevention
• Pregnant women and the immunocompromised should avoid animal contact on farms and unpasteurized milk or milk products, or inadequately cooked meats
• High-risk foods, e.g. soft cheese and pâté, should be stored below 4°C

• Salads should not be eaten beyond their 'use by' date
• Dairy products should be monitored for Listeria if pasteurization is not possible.

Viral meningitis
This is the commonest form of meningitis worldwide and may be caused by a wide range of viruses.

Epidemiology
• Coxsackie- and echoviruses of the enterovirus group:
  • are the commonest cause of meningitis globally
  • occur mainly during summer and early autumn
  • occur both sporadically and in outbreaks, prevalence of individual serotypes varying markedly in time and place
  • are commoner in children and young adults
• Mumps meningitis:
  • is the second commonest form in countries without routine childhood measles, mumps and rubella (MMR) vaccine use but rare in countries using the vaccine
  • is commoner in winter and early spring in endemic countries
• Arboviruses are an important cause of meningitis in countries where they are prevalent
• Acute HIV seroconversion illness may present as meningitis
• Herpes zoster (rarely clinically apparent) and HSV-2 infection are other relatively common causes
• Poliovirus may present only as viral meningitis
• Lymphocytic choriomeningitis (LCM), an arenavirus transmitted through contact with infected rodent urine, is now almost extinct in the UK but occurs infrequently in other areas.

Clinical features
• The onset is acute with fever, headache, photophobia, vomiting and signs of meningeal irritation
• A faint pink, maculopapular rash (often rubella-like) may be present with some enteroviral infections

- Presence of parotitis is usual in mumps meningitis but may be absent.
- Headache in viral meningitis may be intense and may persist for a week even after subsidence of fever and neck stiffness.

Diagnosis
- Viral meningitis is unlikely if the patient looks toxic and sick, otherwise distinction from bacterial meningitis requires CSF examination
- WCCt is usually normal or slightly lymphopenic. Significant neutrophilic leucocytosis and raised ESR or plasma viscosity should suggest bacterial infection
- CSF shows typical changes of viral meningitis (see Table 7.1)
- Differentiation is from:
  - early tuberculous meningitis
  - listeria meningitis (CSF may be lymphocytic)
  - bacterial meningitis where the CSF picture has become modified from preadmission antibiotic therapy (this is rare; usually cytology and biochemistry remain fairly suggestive, though organism may not be found)
  - brain abscesses and other forms of parameningeal sepsis
  - fungal, leptospiral, syphilitic and Lyme meningitis.

Treatment and prognosis
Recovery is usually complete within 10 days.

Poliovirus infection and poliomyelitis
Epidemiology and pathogenesis
- Poliovirus, an enterovirus, causes a spectrum of illnesses, ranging from inapparent infection to acute flaccid paralysis (poliomyelitis)
- It exists in three antigenically distinct types — 1, 2 and 3. Type 1 is the most virulent. Only humans are infected
- Global eradication of this ancient and dreaded infection appears to be imminent:
  - Since the initiation of the global poliomyelitis eradication initiative in 1988, the number of endemic countries has decreased from 125 to 10, and the number of polio cases decreased > 99%, from an estimated 350 000 to < 1000
- Of the six WHO regions, the Americas, Western Pacific and European regions are now officially polio free
- Significant polio transmission now exists primarily in northern India, Afghanistan, Pakistan, Nigeria, Democratic Republic of Congo and Angola
- Transmission is expected to cease in these areas in the near future
- A 3-year consolidation period will follow, after which global eradication will be certified
- Transmission is faecal–oral, usually through direct contact with another infected person, rarely via indirect food or fomite spread
- The IP is 7–14 days
- Paralysis results from damage to motor nuclei in the spinal cord and/or brain.

Clinical features
The outcome of infection may be:
- asymptomatic (most common)
- a short non-specific febrile illness
- viral meningitis
- paralytic illness, characterized by pure motor weakness. This occurs only in a small minority of those infected.

Faecal virus excretion continues for several weeks after fever subsides.

Complications
- Respiratory failure occurs from paralysis of the muscles of respiration or brainstem involvement
- Recovery of muscle power is rarely complete and, in severe cases, extensive paralysis persists, leading to atrophy and deformity.

Diagnosis and treatment
- All cases of acute flaccid paralysis require thorough investigation to establish cause (see Table 7.3)
- Complete bed rest, control of fever and pain during the early stage; ventilatory support if respiratory failure develops and rehabilitation once acute stage is over.

Prevention

• Success of polio eradication has depended on saturation coverage of communities (by supplementing universal childhood vaccination with national and subnational vaccination days and additional mop-up campaigns) with attenuated live oral polio vaccine (OPV), which interrupts transmission of wild poliovirus in the area
• OPV vaccine colonizes the gut, producing local as well as humoral immunity, and also spreads to and protects susceptible contacts
• OPV remains part of the childhood immunization schedule in the UK and many other countries
• An inactivated poliovirus vaccine (IPV) also give good immunity and is used in the USA
• OPV will be phased out once transmission of wild virus ceases, to be replaced by IPV everywhere during the 3-year consolidation period.

Other enteroviral syndromes

• Enteroviruses belong to the Picornaviridae family (RNA viruses) and, apart from polio, other recognized human enteroviruses are coxsackieviruses (24 group A and six group B serotypes), echoviruses (34 serotypes) and 'new' enteroviruses (types 68–71)
• A wide range of clinical syndromes are caused by these agents
• They all infect the human gut and viraemia is common

Epidemiology

• Occurrence is worldwide—sporadic and epidemic
• Transmission is faecal–oral
• The IP is usually 2–5 days
• Asymptomatic infections are very common.

Clinical syndromes

• *Acute encephalitis* (meningitis much more common)
• *Bornholm disease* (epidemic myalgia, pleurodynia):
  • caused by group B coxsackieviruses
  • characterized by severe lower intercostal or upper abdominal muscular pains, associated with fever and malaise

• pain fluctuates in intensity and may be pleuritic
  • symptoms usually subside within 1 week but may relapse
• *Simple febrile illness*, or undifferentiated upper respiratory tract infections (coxsackie- and echoviruses)
• *Herpangina:* painful tiny greyish ulcers over the soft palate and fauces (coxsackie A)
• *Fever with maculopapular rash* (coxsackie- and echoviruses)
• *Hand–foot–mouth disease,* characterized by vesicles on hands and feet and ulcers in the mouth in young children (mostly coxsackie A16)
• *Pericarditis or myocarditis:* occur rarely in adults and teenagers (coxsackie- and echoviruses)
• *Neonatal infection:* severe disseminated infection with myocarditis, hepatitis and encephalitis (mainly echovirus)
• *Acute flaccid paralysis:* similar to poliomyelitis (coxsackie and echoviruses)
• Enteroviral haemorrhagic conjunctivitis:
  • is usually caused by enterovirus 70 and occasionally by a variant of coxsackie A24
  • enterovirus 70 has caused large outbreaks of conjunctivitis in many areas of India, southeast Asia, Africa, Central and South America
  • transmitted by hands and contaminated towels
  • symptoms appear within 2 days of exposure to an infected person
  • subconjunctival haemorrhage is common
  • recovery is usual within 10 days.

Diagnosis and treatment

• This requires isolation of the virus from faeces, and less commonly from throat swab or CSF (or from conjunctiva if eyes are involved); or PCR
• Serology can be useful to confirm diagnosis for some syndromes
• There is no effective treatment.

Viral encephalitis

Encephalitis is a diffuse inflammatory condition of the brain and can be caused by a wide range of viruses (see p. 73). Two pathologically distinct forms of acute encephalitis are recognized: primary and postinfective.

Primary encephalitis

This is due to direct virus attack on the brain, as caused by herpes simplex virus (HSV), enteroviruses, rabies, arboviruses, HIV, influenza, EBV and CMV. There is neuronal inflammation with perivascular infiltration of inflammatory cells. The grey matter is predominantly involved.

Epidemiology

In the UK:
• primary encephalitis is quite uncommon and usually sporadic
• HSV is responsible for the majority of the severe cases in the adult (most result from reactivation of virus dormant in the central nervous system), and has a predilection to attack temporal lobes and cause necrotizing encephalitis
• coxsackie- and echoviruses, and primary HIV infection account for a number of others
• rarer causes include VZV, EBV, adenoviruses, influenza virus and CMV (in the immunocompromised)
Globally, arboviruses are an important cause (very rare in UK); of note are:

West Nile virus

• Has been described in Africa, the Middle East, Europe, west and central Asia, Oceania and, since 1999, the USA with rapid spread (nearly 4000 cases reported in 2002 with about 5% case fatality rate)
• Is spread by infected culex mosquito bites and apart from people also infects animals (e.g. horses) and birds (mainly crows), which are the reservoirs of infection
• Transmission via blood transfusion, organ transplantation and maternal–fetal transmission has also been noted in USA
• The IP is 3–14 days
• Most infections are inapparent or mild febrile illness
• Approximately 1/150 will have neurological involvement (encephalitis more than meningitis,

rarely polio-like illness and Guillain–Barré syndrome).

Japanese B encephalitis (JE)

• Is the most important cause of viral encephalitis in south-east Asia and the western Pacific
• >50000 cases occur annually with a case fatality rate of around 25%
• It is prevalent in rural and agricultural areas where pigs and birds act as animal hosts and mosquitoes transmit the virus from viraemic animals to humans
• The IP is usually 4–14 days
• Most infections are inapparent or mild febrile illness; about 0.2% develop encephalitis which tends to be severe.

Tick-borne encephalitis (TBE)

• Occurs seasonally during spring and early summer in the forested areas of Scandinavia, eastern Europe and northern Russia; a related one (Powassan encephalitis) occurs in Canada and the USA
• Ticks are the vectors and small mammals are the vertebrate hosts
• Human cases are related to outdoor activities in forested areas or drinking of raw goat's milk
• The IP is usually 4–17 days
• Most infections are inapparent.

Other viruses of importance

• Rabies (see p. 226)
• Enterovirus 71 (causes epidemic encephalitis).

Postinfective meningoencephalitis (acute disseminated encephalomyelitis, ADEM)

• Measles, chickenpox and rubella viruses can also cause encephalitis but pathogenesis is different from primary encephalitis
• An immune-related inflammatory response, triggered by virus or other antigen, is thought to be responsible. There is widespread demyelination of the white matter of the brain and often of the spinal cord
• A similar illness used to occur following ani-

mal brain-derived rabies vaccines and smallpox vaccine
• ADEM may develop following nondescript respiratory illnesses or without any apparent preceding clinical event.

Clinical features of acute encephalitis
• The onset is acute, with fever, headache, vomiting, irritability and photophobia
• Meningism may be present (especially in arboviral encephalitis)
• Consciousness level may fluctuate or patient may develop confused and abnormal behaviour. Later, the patient becomes progressively drowsy and even comatose
• Involuntary movements, twitching or frank convulsion may appear (rare in enteroviral meningitis)
• In HSV encephalitis, focal signs such as hemiplegia or aphasia are common, but are unusual in other types
• Myelitis may develop in ADEM.

Diagnosis
• The clinical picture and CSF changes (see Table 7.2) are usually sufficient to distinguish encephalitis from the many toxic and metabolic conditions that may mimic it but CT scan is necessary to exclude intracranial abscess
• Tuberculous meningitis must be excluded
• Virus isolation should be attempted from throat swab, faeces, urine and CSF in all cases of primary encephalitis
• HSV, enteroviruses and VZV can be efficiently diagnosed by PCR of CSF
• Detection of IgM antibody to West Nile virus in serum or CSF is diagnostic
• Otherwise, paired samples of sera should be collected for serological studies against a wide range of possible viral agents and also for mycoplasma which can cause encephalitis
• MR scans help in distinguishing ADEM from primary encephalitis.

Treatment
• Specific treatment is only available for HSV encephalitis and should be begun in all acute sporadic encephalitis whilst investigations are continuing
• Otherwise treatment of encephalitis is entirely supportive, with maintenance of vital functions and measures to reduce raised ICP (dexamethasone)
• IV acyclovir reduces mortality and morbidity in HSV encephalitis
• High-dose corticosteroids may be beneficial in ADEM.

Prognosis
• Mortality varies from negligible (enteroviral encephalitis) to high (JE) and 100% (rabies)
• Neuropsychiatric sequelae are not uncommon in survivors of severe encephalitis.

Acute polyneuritis (polyradiculitis, Guillain–Barré syndrome)
Epidemiology and pathogenesis
• Occurs sporadically throughout the world and is now the commonest form of acute flaccid paralysis in the developed world
• Infection-triggered immune-mediated damage to the myelin covering of the peripheral nerves appears to be the underlying pathogenesis
• Often, a preceding infection cannot be proven
• *Campylobacter jejuni* is an important cause; EBV and CMV account for a small number of cases
• Several cases of West Nile virus infections in the USA have developed Guillain–Barré syndrome
• Smallpox vaccines and animal brain-derived nerve tissue vaccines (e.g. for Japanese B encephalitis and tick-borne encephalitis; previously older rabies vaccines) can rarely cause a similar illness
• History of a preceding upper respiratory infection or sore throat may be found in about 25% of patients.

Clinical features and diagnosis
• There is progressive symmetrical lower motor neurone-type paralysis of the limbs and trunk over several days, with symmetrical distal sensory loss and areflexia

• Paralysis of respiratory and facial muscles and bulbar paralysis develop in severe cases
• The differentiation is from diphtheritic polyneuritis, poliomyelitis, botulism and acute myelitis (see Table 7.3)
• The CSF shows greatly raised protein without any pleocytosis, and a normal glucose.

Prognosis and treatment
• Most cases recover fully, though this may take a long time
• Occasional deaths are from complications of intensive therapy. A few patients have spinal cord involvement (myeloradiculitis) and suffer from permanent residual weakness
• Plasma exchange or IV immunoglobulin helps to speed up recovery
• Respiratory failure requires management in an intensive care unit with assisted ventilation and establishment of airway
• Corticosteroids are ineffective.

Botulism
Epidemiology and pathogenesis
• This is a paralytic illness caused by the neurotoxin of *Clostridium botulinum* which blocks the release of acetylcholine at the neuromuscular junction
• The organism is an anaerobic, spore-forming bacterium which is widespread in soil
• The spores are heat resistant and can survive up to 2 h at 100°C but are killed at 120°C
• Most cases of human botulism are due to ingestion of toxin via food in which contaminating spores have been allowed to germinate in an anaerobic environment, thus producing the toxin
• Consumption of preserved or canned foods such as meat, fish, fruit and vegetables which have been inadequately heat treated is often implicated
• Botulism may be caused by contamination of wounds by *C. botulinum* with subsequent production of exotoxin
• Rarely, infants may swallow spores which later germinate and produce toxin (infant botulism). Toxigenesis cannot occur in mature adult guts

• Food-borne and infant botulism are very rare in the UK but a small number of cases of wound botulism occur each year in injecting drug users
• It is somewhat more common in the USA with over 100 cases annually (infant more than food-borne)
• The IP of food-borne botulism is 12–36 h; in infant and wound botulism some time may elapse between spore ingestion/wound contamination and toxigenesis.

Clinical features
• In food-poisoning and wound botulism:
  • patient may present with blurred vision, slurred speech, drooping eyelids, difficulty swallowing and muscle weakness
  • within 24 h, flaccid paralysis of the limb and trunk muscles may develop, leading to respiratory failure
  • sensation remains intact
  • fever is absent
  • recovery is gradual over a period of weeks or months, but complete
• In infant botulism:
  • most cases have a mild and transient illness with constipation and lethargy as prominent features
  • in a few infants, illness is more severe with difficulty in feeding, paralytic squints, flaccid limbs and trunk weakness.

Diagnosis
• This is primarily clinical
• Suspect botulism in any afebrile patient with acute descending flaccid paralysis
• Differential diagnoses include acute polyneuritis, bulbar myasthenia of rapid onset, diphtheria with palatal and extraocular palsies, organophosphorous poisoning, cerebrovascular accidents involving the basilar artery, and paralytic shellfish poisoning
• Confirmation is by demonstrating toxin in the blood or faeces, or the organism in wound specimens.

Treatment and prevention
• Botulinum antitoxin is effective in reducing severity of illness if given early

- Wound and infant botulism should be treated with penicillin or metronidazole to stop toxigenesis
- Ventilatory support if needed
- Modern and efficient commercial standards of canning.

Tetanus
Epidemiology
- Tetanus is caused by *Clostridium tetani*, a Gram-positive, anaerobic, spore-forming bacillus, present in the bowels of many herbivorous animals and widely distributed in soil
- In the absence of active immunization, a person of any age can develop tetanus through wounds contaminated with soil
- Because of routine childhood immunization, tetanus is rare in Britain (less than 20 cases/year) and other developed countries
- Adults aged >60 years are at greatest risk, particularly women, who are likely to have been born before the introduction of childhood immunization and less likely to have received vaccine through military service
- In developing countries tetanus remains an important cause of death
- Neonatal tetanus is a special problem in some developing countries through contamination of the umbilical stump with soil or animal dung (for therapeutic purposes).

Pathogenesis
- In wounds contaminated with spore-bearing soil, anaerobic conditions created by the presence of foreign bodies and devitalized tissues encourage active vegetative growth of *C. tetani*, leading to toxin production (tetanoplasmin)
- The toxin travels proximally along the nerves to reach the nervous system and produces tetanus by two mechanisms:
  by blocking acetylcholine release at the myoneural junction and
  by countering the inhibitory influences on muscle reflex arcs
- This leads to muscle rigidity and spasms. Once 'fixed' in the spinal cord, the toxin can no longer be neutralized by antitoxin

- Removal of inhibitory influences on the autonomic nervous system causes increased autonomic activity, causing tachycardia, sweating and hypertension
- The IP is normally 5–15 days, but may be longer.

Clinical features
- Rigidity stage:
  - trismus (painfully stiff jaw muscles) — often the first symptom
  - difficulty in opening the mouth (lockjaw)
  - dysphagia may develop
  - fever, if present, is mild
  - in 24h, stiffness spreads to neck, back, chest and abdominal wall muscles. Arms and legs are only slightly affected
- Spasmodic stage:
  - generally within 1–2 days, intermittent, spasmodic, painful contractions of the affected muscles develop — often accompanied by pallor and sweating
  - spasms cause grimacing of the face (risus sardonicus), and arching of the neck and back (opisthotonus)
  - spasms of laryngeal and respiratory muscles cause respiratory failure
  - spasms occur spontaneously or may be triggered by noise, coughing and movements
- In severe cases, signs of sympathetic overactivity appear:
  - profuse sweating, fever, hypertension/hypotension, tachycardia, cardiac arrhythmia
- In surviving patients, the spasms gradually subside after 2–3 weeks and muscle rigidity disappears after another 1–2 weeks
- In mild cases, there is often rigidity alone, and this may rarely be localized only to the site of injury.

Diagnosis
- This is usually clinical, although *C. tetani* may occasionally be found in the wound
- Tetanus is extremely rare in an adequately immunized person. Presence of antitoxin levels of 0.01 IU/mL in the admission sample in a patient with tetanus-like spasms excludes the diagnosis.

Differential diagnoses

- Drug-induced acute dystonia (metoclopromide, prochlorperazine) involving head and neck muscles — rigidity is absent and oculogyric spasms are characteristic. The condition responds dramatically to 2 mg IV benztropine
- Trismus may be mimicked by dental abscess, mumps or temporomandibular joint problems
- Rabies, strychnine poisoning and spinal myoclonus.

Treatment

- Patients with muscle spasms require intensive care
- The important steps in management are:
  - administration of human tetanus immunoglobulin 20 000 IU intravenously, followed by debridement of the wound
  - benzylpenicillin IV or IM for 10 days to eradicate the existing foci of infection and stop further toxin production
  - sedation of the patient: diazepam is used widely and may control mild spasms; patients should be nursed in a quiet environment to avoid triggering spasms
  - tracheostomy in patients with spasms: to guard against sudden life-threatening laryngospasm
  - if sedation is not controlling spasms: paralysing drugs and ventilation

Table 7.4 Active and passive immunization against tetanus in injured patients.

- attention to fluid and electrolyte balance and nutrition
- labetalol (has both α- and β-blocking properties) or diazoxide for sympathetic hyperactivity
- Before discharge the patient should be actively immunized as the disease does not produce immunity.

Prevention

- Active immunization with tetanus toxoid gives protection for at least 10 years, probably much longer (no death reported in anyone following a full course of primary vaccination without any further booster)
- Universal childhood immunization is the key to the control of tetanus in any country
- In developing countries, immunization of women at antenatal clinics prevents neonatal tetanus
- Effective wound care:
  - All wounds should be thoroughly cleaned, with removal of foreign bodies and dead tissue. Penicillin or erythromycin prophylaxis for contaminated or infected wounds may reduce the likelihood of tetanus
  - Patients with injuries should be considered for active and passive immunization as detailed in Table 7.4.

Prognosis

Modern management reduces the mortality of severe tetanus from 60% to 10–20%. Most

Tetanus vaccination history	Type of wound	
	Clean, minor wounds < 6 h old	All other wounds
Uncertain, none, incomplete course	Give tetanus toxoid*; consider antibiotics	Give tetanus toxoid* and immunoglobulin†
Three or more doses		
Last dose > 10 years ago	Give tetanus toxoid booster	Give tetanus toxoid booster
Last dose < 10 years ago	Nil	Nil

* Complete basic course of tetanus immunization.
† Human antitetanus immunoglobulin.

deaths are due to sympathetic hyperactivity. Mild or localized tetanus has no mortality.

Prion encephalopathies

- A group of progressive non-inflammatory degenerative diseases of the brain affecting animals and humans are caused by abnormal prions, which are transmissible, although they are not living organisms
- Prions are proteins found in normal cells
- Prions can transform into a pathogenic abnormal form, accumulation of which within nerve cells leads to neurodegeneration
- Abnormal prion protein is very resistant to all common methods of inactivation.

Examples of prion encephalopathies affecting animals

- Scrapie in sheep

- Bovine spongiform encephalopathy (BSE) in cattle
- Transmissible encephalopathy of mink.

Human prion encephalopathies

Creutzfeldt–Jakob disease (CJD)
Four different types are recognized, based on how the illness is caused in an individual.

1 *Sporadic CJD*—caused by transformation of normal prion protein into abnormal form by a chance mutation, numerically the most common form.

2 *Genetic or familial CJD*—abnormal prion is produced by an inherited genetic disorder. It is very rare.

3 *Iatrogenic CJD*—caused by accidental transmission of abnormal prion from another person with the illness, through medical or surgical procedure. Several such cases resulted in the UK via use of human pituitary glands as a source of human growth hormone. It is very rare.

Table 7.5 Differences between sporadic CJD and vCJD.

Features	Sporadic CJD	vCJD
Prevalence	Worldwide, about 1 case/ million	Most cases have occurred in UK (129 by 2002). Outside UK, 6 in France, 1 each in Ireland, Italy and USA
Clinical		
Age of onset	Middle-aged and elderly	Younger individuals, median age 26 years. Often psychiatric or behaviour problems; frank neurological signs develop later
Presentation	Usually with frank neurological signs e.g. dementia, ataxia, pyramidal and extrapyramidal signs, myoclonus	
Duration	Typically short, usually a few months	Often a year or more
Investigations		
EEG	Typically, periodic sharp wave complexes	Often normal or non-specific
MR scan	Often non-specific cortical atrophy	Prominent symmetrical high signal intensity on T2-weighted/proton density-weighted MRI
Prion present in lymphoreticular tissue	No	Yes
Neuropathology	Distinctive	Distinctive
Structure of prion protein	Distinctive	Distinctive

4 *Variant CJD (vCJD).*
  • First reported in the UK in 1996 and is believed to have resulted from transmission of infection from BSE in cattle to humans via contaminated beef.
  • BSE probably originated in the UK in the 1970s — possibly from a single cow or other animal that developed the disease through chance mutation. Subsequently an epidemic of BSE developed through the practice of using cattle offal as cattle feed.
  • Transmission to humans occurred by processed food items that contained infectious bovine tissue such as brain or spinal cord.

Differences between sporadic CJD and vCJD
See Table 7.5.

Treatment
None available.

Prevention
  • Due to the stringent measures undertaken, cattle in the UK are free of BSE and the vCJD risk is deemed to have been eliminated from the food chain
  • Strict guidance is in place to prevent transmission via surgical instruments and via use of human tissues and organ transplants
  • Leucodepleted blood is used in transfusion
  • Non-UK-sourced plasma and blood products are in use
  • Children <6 years are given non-UK-sourced blood transfusion.

Kuru
This has occurred exclusively among the cannibalistic tribal people of New Guinea. A transmissible agent similar to CJD is involved. The disease is characterized by progressively worsening cerebellar symptoms and wasting, ending in death. New transmissions have ceased following cessation of cannibalism.

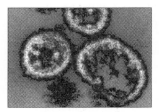

Rashes, and Skin and Soft Tissue Infections

Clinical syndromes

Maculopapular rash and fever
See Table 8.1, p. 91.
Erythematous rash and fever
See Table 8.2, p. 93.
Purpuric or haemorrhagic rash and fever
See Table 8.3, p. 94.
Papulovesicular/pustular rash and fever
See Table 8.4, p. 95.
Papular or plaque-form rash
See Table 8.5, p. 96.

Specific infections

Measles
Epidemiology
• Causative virus belongs to paramyxovirus family (RNA viruses), has a single antigenic type and affects humans only

• It is highly infectious with a household attack rate of >90%
• Occurs worldwide; 777 000 children died in 2000
• In unimmunized communities most children will have measles before adulthood and 2-yearly epidemics are common
• In many developed countries measles has become rare through national childhood vaccination programmes
• >95% vaccine uptake necessary for elimination of measles
• In 2000, only 86 cases (majority imported) reported in USA; Finland had had no autochthonous case since 1996
• Suboptimal vaccine coverage (<90%) allows continuing virus circulation, leading to shift to infections in higher age group where complications may be more frequent and serious
• Immunity after natural infection is lifelong. Maternal antibody protects the infant up to the age of 6 months
• Transmission is via respiratory secretion from infected children through inhalation of

Table 8.1 Distinguisting features and causes of maculopapular rash and fever.

Cause	Distinguishing features		Occurrence
	Rash	Associated features	
Measles	Discrete maculopapules spreading slowly from hairline downwards and becoming confluent	Prominent fever and coryza precede rash for 3–4 days. Koplik's spots on day 2	Worldwide but now rare in the developed countries with high vaccine uptake
Rubella	Discrete maculopapules spread rapidly from hairline downwards to limbs in 24 h, clearing as rash evolves	Absent prodromal fever or mild and short in adults. Tender postcervical glands. Arthritis common in adults	Worldwide but now rare in the developed countries with high vaccine uptake
Parvovirus B19	Erythema on cheeks followed by diffuse lace-like rash on body	Prodrome absent or mild fever. Arthritis in adults	Worldwide
Human herpes virus 6 infection (exanthem subitum)	Maculopapules appearing on trunk as fever settles	Prominent prodromal fever for 3–4 days before rash	Worldwide, almost exclusively in children > 2
Primary HIV infection	Scattered maculopapules on body	Glandular fever-like illness. Rarely meningitis, encephalitis	Worldwide
Enteroviral infections	Scattered maculopapules on body	Fever, myalgia, headache. Other enteroviral syndromes (see p. 82)	Worldwide
Dengue	Widespread macular or maculopapular, often becoming confluent	Severe headache and myalgia, nausea, vomiting	South and south-east Asia, Africa, Mexico, Central and South America
Typhoid/paratyphoid fever	6–10 maculopapules on lower chest/upper abdomen on day 7–10 of fever	Sustained fever, splenomegaly	Residence in or travel to south Asia, South America, Africa, Middle East, south and east Europe

Continued on p. 92.

Table 8.1 (continued)

Epidemic typhus	Maculopapules on trunk and face and limbs, sparing palms and soles. May become petechial	3–5 days of fever, chill and toxaemia preceding rash	In unhygienic conditions of war and famine in colder climates of Asia, Central Africa, Central and South America
Endemic typhus	Maculopapules on body sparing palms and soles	Similar to endemic typhus but milder	Worldwide in conditions where people live in rat-infested buildings
Scrub typhus	Diffuse maculopapules on trunk spreading to limbs	Fever preceding rash. Eschar at the site of mite attachment	Central eastern and south-east Asia, Himalayas, Northern Australia
Tick-borne spotted fevers	Maculopapules on limbs including palms and soles, later in other areas, may become petechial	Fever, myalgia, headache. Eschar at the site of tick bite	India, Africa, Queensland, Mediterranean region, Siberia
Rocky Mountain spotted fever	Maculopapules on limbs including palms and soles, later spreading to other areas, becoming petechial	Preceding high fever of 2–3 days. Chill, headache, myalgia	USA (rare in Rocky Mountain region), Canada, Mexico, Central America, Brazil
Leptospirosis	Maculopapules, may become petechial	Severe myalgia, fever, injected conjunctiva. Jaundice and renal failure	Worldwide Occupational/recreational contact with rat urine
Rat-bite fever	Red plaques on trunk (Spirillium minor) or maculopapules on limbs (Streptobacillus moniliformis). May be petechial	Prodrome of fever, headache and chills. History of rat bite. Wound flares up at fever onset (Sp. minus)	Rare in West. Sp. minus commoner in Asia

NB. Drug hypersensitivity, systemic lupus erythematosus and Still's disease are non-infective causes of maculopapular rash and fever.

Table 8.2 Distinguishing features and causes of erythematous rash and fever.

Cause	Distinguishing features		Occurrence
	Rash	Associated features	
Scarlet fever	Diffuse punctate erythema on body, sparing circumoral area	Exudative tonsillitis. Strawberry or raspberry tongue. Desquamation	Worldwide. Mostly in children
Toxic shock syndrome (staphylococcal and streptococcal)	Erythematous sunburn-like rash on trunk and limbs	Toxicity and hypotension. Multiorgan dysfunction. Necrotizing fasciitis (in streptococcal variety)	Worldwide. Menstruating women Surgery, burns
Staphylococcal scalded skin syndrome (SSSS) Toxic epidermal necrolysis (drug related — TEN)	Widespread erythematous patches, skin separating to leave raw area. Bullae may form	Nicolsky's sign (see p. 108) Desquamation later. Skin biopsy discriminates between the two	Worldwide. SSSS mostly in children TEN in adults
Erythema multiforme	Papuloerythematous lesions symmetrically on limbs, may affect trunk and face. Target lesions (raised peripheral ring & pale centre). Bullae may form	Nicolsky's sign. Mucosa of mouth, eye or urethra may be affected	Worldwide, mainly children and adults, as an immunological reaction to drugs (e.g. sulphonamide) or infections (e.g. *Mycoplasma*, HSV)
Kawasaki disease	Erythematous oedema of limbs, erythematous rash on trunk Desquamation of fingertips later	Spiking fever for > 5 days. Cervical adenitis, dry lips, injected eyes Raised neutrophils and platelets Coronary artery aneurysm	Mostly reported in the developed countries, highest incidence in Japan Mostly in young children ?Infection-triggered Immune-related vasculitis
Erythema marginatum	Fluctuating circinate erythematous lesions with a clear centre and a serpiginious outline	Recent pharyngitis. Signs and symptoms of rheumatic fever	Worldwide but commoner in developing countries
Erythema nodosum	Tender, shiny red areas on shins, less commonly on forearms or elsewhere May last for several weeks	Fever, polyarthralgia	Worldwide Immune-related to infections, drugs, sarcoid

NB. Drug hypersensitivity can cause an erythematous rash and fever.

| Cause | Distinguishing features | | |
	Rash	Associated features	Occurrence
Meningococcaemia	Rapidly developing purpuric or petechial rash	Meningitis Hypotension	Worldwide
Viral haemorrhagic fevers	Widespread purpuric rash	Shock, visceral bleeding	Residence or travel to endemic countries
Henoch–Schonlein purpura	Petechiae on distal limbs and buttocks	Arthralgia, abdominal pain Haematuria infrequently Evidence of recent *S. pyogenes* infection	Worldwide Commoner in children and young adults

NB. Thrombocytopenia from any cause may present with purpura. Infection-related causes are rubella, HIV, haemolytic–uraemic syndrome and septicaemia-related coagulopathy. Rashes of enteroviral, rickettsial, leptospiral and rat-bite fever infections may also be petechial.

Table 8.3 Distinguishing features and causes of purpuric rash.

airborne droplets or by direct contact. Infectivity persists from just before the prodromal period until the 4th day of the rash
• Incubation period (IP) is about 10 days (range 7–18 days).

Pathogenesis
• Viraemia follows invasion of respiratory epithelium
• The rash is probably immune mediated, coinciding with antibody formation
• Encephalitis is thought to result from local immunological reaction to virus-infected cells
• After recovery both humoral and cellular immunities develop.

Clinical features
• Prodromal illness:
  • 3–4 days of fever, coryza and conjunctivitis, dry cough
  • Koplik's spots (tiny white spots on red macules on the buccal mucosa, opposite the molar teeth) appear on day 2 or 3
• Rash:
  • from day 4, dusky red discrete maculopapules spread from hairline downwards to reach extremities in 2–4 days, becoming confluent, fading later, leaving a fine desquamation and often staining.

Complications
• Bacterial otitis media (1/20), bacterial pneumonia (1/25), febrile convulsion (1/200)
• Postinfective encephalitis (1/1000), late-onset subacute sclerosing panencephalitis (1/100 000), progressive encephalitis in the immunocompromised
• Severe measles in malnourished children in the tropics: prolonged rash, diarrhoea and secondary bacterial complications.

Diagnosis
• Laboratory confirmation is recommended for atypical cases and in areas where measles has become rare
• Demonstration of specific IgM antibody in blood or saliva, or fourfold rise of IgG antibody are confirmatory
• Detection of antigen in nasopharyngeal secretion is the best means of diagnosis in the immunocompromised as antibody response may be defective.

Treatment
• Children should be kept away from school until after 4 days from rash onset

	Distinguishing features		
Cause	Rash	Associated features	Occurrence
Chickenpox (varicella)	Erythematous papules progressing to vesicles, pustules and crusts, appearing in successive crops on face and trunk, sparsely on distal limbs	Prodrome absent or mild malaise Ulcerating vesicles in mouth	Worldwide Predominantly a disease of children
Disseminated herpes zoster	Similar to chickenpox A zoster rash is usually present but may be absent	Immunodeficiency	Worldwide
Smallpox	Centrifugally distributed papules progressing to vesicles, pustules and crusts at the same time. Palms and soles affected	Prodromal fever of 3–4 days. Patient toxic and ill	Does not exist in nature Potential for bioterrorism attack
Eczema herpeticum	Vesicles and crusts on eczematous skin, may spread to healthy areas	Eczema	Worldwide
Hand, foot and mouth disease	Tiny clear vesicles on dorsum of hand and feet	Discrete mouth ulcers Mild fever	Worldwide Affects young children
Rickettsia pox	Papulovesicles on trunk, limbs and face	Preceding eschar where bitten by mite Fever, regional lymphadenopathy	Rare infection in USA, Korea, Russia
Ecthyma gangrenosum	Plaques progressing to pustules or bullae, then necrotic crusts, commoner in axillae and groins.	Neutropenic patients Evidence of pseudomonas septicaemia	Worldwide

Table 8.4 Distinguishing features and causes of papulovesicular/pustular rash and fever.

- In hospital single-room isolation is necessary
- Most cases require symptomatic care only. Co-amoxiclav is appropriate for secondary bacterial infections.

Prevention
Vaccination:
- In Britain and many other countries, live attenuated measles vaccine is given routinely (as measles–mumps–rubella, MMR) during the 2nd year of life with a subsequent booster dose

- Vaccination gives lasting protection in 90% following a single dose and >99% with a second dose
- About 5% of recipients develop a short-lasting febrile illness, sometimes with a transient rash after vaccination, but other side-effects are very rare
- Fears of a link with autism and Crohn's disease are unfounded
- It is safe to be given to HIV-positive infants but contraindicated in other forms of immunosuppression, in pregnancy and in those with severe egg allergy

	Distinguishing features		
Cause	Rash	Associated features	Occurrence
Pityriasis rosea	Oval, red-brown papules with fine scales on trunk, following slopes of ribs giving Christmas tree-shaped distribution on back	Solitary round lesion anywhere on body precedes rash (herald patch) Fever usually absent	Worldwide, mainly in young adults
Seondary syphilis	Round, coppery papules, with fine scales symmetrically distributed on body, prominent on palms and soles	Fever, malaise Mucus patches, condyloma lata, alopecia. History of chancre in some	Worldwide in sexually active adults

Table 8.5 Distinguishing features and causes of papular or plaque-form rash.

- Vaccine virus does not spread to others
- In institutional outbreaks, vaccines given to contacts within 72 h of exposure, or normal immunoglobulin given within 6 days, will prevent or attenuate illness.

Prognosis
- Uncomplicated measles is rarely fatal in previously healthy children
- Pneumonia and infrequently encephalitis cause most fatalities
- It is an important cause of morbidity and mortality in malnourished children in developing countries.

Rubella (German measles)
Epidemiology
- Rubella virus is a togavirus (RNA virus) and has a single antigenic strain. Humans are the only hosts
- It is highly infectious and has a household attack rate of around 80%. In unvaccinated communities around 80% of adults would have had prior infections. Inapparent infections are common
- Occurs worldwide but now rare in countries with efficient national childhood vaccination campaigns
- Transmission is via airborne droplets of nasopharyngeal secretion of infected people 5 days before to 5 days after their rash develops.

Transplacentally infected babies are often chronically infected, excreting virus in throat and urine (for up to 1 year) and are potential sources of infection to others
- Immunity is lifelong
- The IP is 16–18 days.

Pathogenesis
- Viraemia follows invasion of respiratory mucosa
- The rash is immune related, coinciding with the development of antibodies at the end of viraemia
- Fetal damage probably results from arrest of cellular mitosis of unknown mechanism.

Clinical features
- Prodromal illness:
  - usually absent in young children but adults often have a short fever
  - enlarged, tender postcervical glands
- Rash: discrete, rose-pink macules spread from hairline to trunk and limbs within 24 h then rapidly fade
- Polyarthralgia: may affect finger joints and other large joints bilaterally in adults, lasting over several weeks.

Complications and prognosis
Rubella is usually very benign in the acquired form, except for two rare complications — postinfectious encephalitis (1/5000) and immune thrombocytopenia (1/3000).

In infection during early pregnancy:
- Fetal death may occur or
- Baby may have *congenital rubella syndrome*:
  - baby may be born with permanent defects, i.e. cataract, microcephaly, perceptive deafness, patent ductus arteriosus or septal defect, mental retardation or diabetes
  - low birth weight, thrombocytopenia, jaundice, hepatosplenomegaly or pneumonia may be present
- The risks and severity of fetal damage are related to time of infection:
  - highest incidence (up to 80%) and major defects during the first 4 weeks of gestation
  - 25% during 8–12 weeks
  - around 10% during 12–16 weeks (mostly deafness, detected only later due to child's learning difficulty)
  - rare after 20 weeks.

Diagnosis
- Clinical diagnosis is unsatisfactory as rashes due to parvovirus B19 and enteroviruses are often rubelleiform
- Serological confirmation (presence of rubella-specific IgM in blood or saliva or fourfold rise of IgG antibody during convalescence) is mandatory during early pregnancy and also advisable in suspected rubella in low-incidence countries for epidemiological reasons
- Congenital rubella syndrome is diagnosed by demonstration of IgM antibody and/or virus isolation from throat or urine.

Treatment
- There is no specific treatment
- Termination of pregnancy may have to be considered in maternal rubella, depending on the stage of pregnancy and maternal wish.

Prevention
- Normal immunoglobulin following exposure does not prevent infection
- In industrialized countries children are now routinely given highly effective (>95% seroconversion) live rubella vaccine (as MMR) during their 2nd year of life. Immunity is long-lasting
- The vaccine is contraindicated in the immunocompromised and during pregnancy (for fear of fetal damage although none has been reported)
- Vaccine virus may appear in the nasopharynx following vaccination but does not spread to others.

Parvovirus B19 infection (erythema infectiosum, slapped cheek syndrome)
Epidemiology
- The virus belongs to the Parvoviridae family (DNA viruses) and has a single antigenic type. Only humans are affected
- Occurrence is worldwide—both sporadically and in epidemics. Household attack rate is around 50% and lower in other institutional settings
- Infection is common in children and is often asymptomatic
- Immunity is lifelong
- Transmission is through inhaled droplets or contact with infected respiratory secretions, or transplacental in the case of fetal infection; rarely through blood products. Once rash develops infectivity ceases
- The IP is 13–18 days (range 4–20 days).

Pathogenesis
- Viraemia follows inoculation of respiratory mucosa
- The rash coincides with the development of specific antibodies and is immune mediated
- The virus binds to red cell precursors and shuts off erythropoiesis for 7–10 days
- In normal individuals this does not cause any problem because of the long red cell survival
- In those with short red cell survival (e.g. haemolytic anaemia) transient aplastic crisis may develop
- Fetal death and hydrops fetalis occur through the same mechanism
- In the immunocompromised, red cells may be chronically infected, leading to severe anaemia.

Clinical features
- Striking erythema of both cheeks (slapped cheek appearance) is often the first sign
- 1 or 2 days later a maculopapular rash appears on the trunk and limbs which often assumes a lace-like appearance. This may reappear after

fading, particularly on exposure to sunlight. In adults, the rash is often rubella-like
• Fever is unusual
• Symmetrical arthralgia/arthritis of peripheral joints, often without rash, is often the presentation in adults. This usually resolves in 3 weeks.

Complications
• Transient aplastic crisis (in patients with haemolytic anaemia) and severe chronic anaemia (in the immunocompromised)
• Infection in pregnancy can rarely cause hydrops fetalis and fetal death; asymptomatic fetal infections are commoner
• Congenital anomalies do not occur
• Risk of fetal loss and hydrops fetalis is between 3 and 9% in infections during the first 20 weeks of gestation, and appears to be the same whether infection is symptomatic or not.

Diagnosis, treatment and prevention
• Laboratory confirmation is necessary in pregnant women or when haematogenous, rheumatological or fetal complications are suspected by demonstrating:
   • virus-specific IgM in blood (remains positive for up to 2 months) or
   • viral DNA in serum from anaemic patients or in aborted fetal tissue
• Virus-specific IgG denotes past infection
• In confirmed infection in pregnancy, serial ultrasound is performed to detect hydrops fetalis. Intrauterine fetal transfusion improves survival
• Therapeutic termination is not generally indicated, neither is routine antenatal screening for maternal antibodies
• Intravenous normal immunoglobulin has been found useful in treating chronic infections
• In hospital outbreaks susceptible persons with haemolytic anaemia or immunodeficiency or early pregnancy should be excluded from the affected wards. High-risk patients should be cared for by B19 IgG-positive staff
• In school or workplace outbreaks, parents and employees should be advised of the risks of transmitting and acquiring infection and who is vulnerable, so that they can decide whether to avoid the affected area

• A vaccine is at an early stage of development. The efficacy of normal immunoglobulin in postexposure prophylaxis has not been assessed.

Herpes simplex virus infections
Epidemiology
• Herpes simplex virus (HSV) belongs to the family of Herpesviridae (DNA virus) and affects only humans. Two types exist—HSV-1 and HSV-2
• Infections occur worldwide. Around 10% of primary infections are symptomatic
• Most HSV-1 infections occur during the preschool years and between 70 and 90% of the world adult population have antibodies in their blood from past infection
• HSV-2 infection usually occurs during sexual activity: 20% of adults have antibodies; this figure is higher in sexually promiscuous individuals and lower socioeconomic groups
• Transmission is through direct contact of broken skin or mucosa with orogenital secretions of infected persons
• The IP is around 5 days (range 2–12 days).

Pathogenesis
• After initial multiplication at the site of inoculation the virus travels along nerve endings to regional ganglia for further multiplication
• Thereafter it descends along the nerves to affect larger skin or mucosal surfaces. Contiguous spread to adjacent areas also occurs
• Affected cells show ballooning degeneration with characteristic giant cells and intranuclear inclusions
• As the primary infection subsides, humoral and cellular immunities form but the virus remains dormant in the ganglion cells and may reactivate later, leading to recurrent local disease or asymptomatic virus shedding. Intercurrent illness, sunlight and trauma are known reactivating factors but the cause of reactivation often remains unknown.

Clinical features
Gingivostomatitis
• Commonest clinical manifestation of primary

HSV-1 infection, some cases are caused by HSV-2. Preschool children mostly affected
• Presents with fever, salivation, irritability and food refusal. Examination reveals multiple shallow painful ulcers on tongue, gums and buccal mucosa with vesicles on and around lips. Tonsillopharyngitis is less common. Cervical glands are enlarged and tender. Symptoms subside in 5–7 days.

Cold sores
Usually affect adolescents and adults. Most are due to reactivation of HSV-1 in trigeminal ganglia.
• Cluster of vesicles develop on lip and around mouth, lasting a few days
• A prodrome of burning, tingling and itching for several hours is common
• Mouth ulcers are rare unless patient has cellular immunodeficiency when protracted mouth ulcers and vesicles around mouth are common.

Keratoconjunctivitis
Primary infection may result in unilateral conjunctivitis with vesicles on eyelids. Keratitis may develop with typical dendritic ulcers. A recurrence is milder but keratitis may be troublesome.

Genital herpes
Usually affects sexually active adults. HSV-2 accounts for the majority of cases. A significant minority is now caused by HSV-1 (through orogenital sexual activity).
• Primary infections are often inapparent or mild and undiagnosed; less commonly symptoms are severe with vesicles and ulcers on vulva, vaginal mucosa and cervix, or penis, associated with fever, headache, myalgia, dysuria and discharge. Lesions may spread to adjacent areas. Proctitis may develop
• Healing occurs in 2–4 weeks
• Recurrences are mostly with HSV-2 infections. Symptoms are milder, less extensive and short lived.

Skin herpes
• Intact skin is impervious to HSV but a breach

may be the site of primary infection, resulting in a cluster of vesicles with pain and oedema. There may be recurrences in the same area. Special forms:
• herptic whitlow (site of infection — fingertip)
• herpes gladiatorum (skin herpes in a wrestler through trauma during bout)
• eczema herpeticum — primary infection of eczematous skin may lead to extensive diseases with vesicles and crusts covering a large area with marked associated systemic symptoms. May be fatal if not treated promptly.

Neonatal herpes
HSV can infect a neonate either *in utero* or during birth. Most cases are due to HSV-2.
• Risk of transmission is high (30–50%) when mother develops primary genital herpes near the time of delivery and low (<1%) during the 1st trimester or where she has recurrent genital herpes at term
• Infected neonates often have severe disseminated disease with central nervous system (CNS) and other visceral involvement. Vesicles on skin are common
• Mortality in untreated cases is around 65%.

Neurological syndromes caused by HSV
• Acute sporadic viral encephalitis of adults
• Aseptic meningitis (usually HSV-2) which may be recurrent (Mollaret's syndrome)
• Transverse myelitis, sacral radiculopathy (numbness and tingling of buttocks and perineum, bladder and bowel symptoms) may complicate genital herpes
• Bell's palsy and other cranial neuritis.

Disseminated and visceral herpes
• Occurs in neonates and cellular immune deficiency
• Mortality is high.

Diagnosis
• Clinical diagnosis is sufficient in most cases of childhood gingivostomatitis and cold sores
• Virological diagnosis is important for herpes encephalitis and meningitis, neonatal herpes,

disseminated herpes, genital herpes (knowledge of serotype will indicate the chances of recurrence, i.e. HSV-1 rarely recurs) and skin herpes (distinction from zoster and pyogenic paronychia)

- Virus culture or antigen detection
- Polymerase chain reaction (PCR) for viral DNA in cerebrospinal fluid (CSF) is useful for detecting CNS infection
- Virus-specific IgM in neonatal herpes
- Fourfold rise in IgG antibodies for confirming clinical diagnosis of genital herpes when virus culture is negative
- Virus can be typed serologically.

Treatment

- Intravenous acyclovir is necessary for neurological, neonatal, visceral, disseminated and severe mucocutaneous infections
- Milder mucocutaneous cases can be treated orally with acyclovir, famciclovir or valaciclovir
- Recurrent cold sores may benefit from topical acyclovir if applied very early
- Acyclovir, famciclovir or valaciclovir may be used long term for suppressing genital herpes or for shorter periods to cover triggering factors in cold sores (e.g. exposure to strong sun)
- Acyclovir-resistant HSV (found increasingly in HIV patients) needs treatment with foscarnet or cidofovir
- Ocular infections are treated with topical acyclovir (systemic administration necessary in primary infection).

Prevention

- Patients and carers with active lesions should cover these or avoid vulnerable individuals
- Prevention of neonatal herpes:
  - acquisition of genital HSV infection during late pregnancy is prevented by avoiding orogenital contact with known or suspected genital or orofacial herpes
  - women without genital herpes can deliver vaginally
  - Caesarian section is performed if mother has active genital herpes at term
  - oral acyclovir suppressive therapy during

late pregnancy in women with a history of recurrent genital herpes is controversial.

Chickenpox (varicella)
Epidemiology and pathogenesis

- Chickenpox results from primary infection by varicella zoster virus (VZV), a member of the Herpesviridae family and a strict human pathogen.
  - It is primarily a childhood disease of worldwide prevalence
  - It is highly infectious with a household attack rate approaching 90% (in urban communities 90% of adults have had chickenpox)
  - But it spreads less easily outside, and in isolated rural communities, particularly in the tropics, many adults may remain susceptible
  - Its incidence has declined dramatically in the USA and other countries through routine childhood vaccination
  - Transmission is through inoculation of respiratory tract by infected respiratory secretion or vesicle fluid (chickenpox or zoster) via inhalation or direct contact
- Immunity to chickenpox is lifelong
- The IP is usually 13–17 days (range 10–21 days)
- Viraemia follows nasopharyngeal inoculation leading to:
  - localization in skin causing balloon degeneration of cells with formation of multinucleated giant cells and intranuclear inclusion
  - infection of sensory nerve ganglia where the virus lies dormant after recovery and may reactivate in later life, presenting as herpes zoster (HZ).

Clinical features

- *Prodromal period:* usually absent in children, but adults may have fever and malaise for 1–2 days
- *Rash:*
  - erythematous maculopapules appear on face and trunk and progress through vesicular, pustular and crusting stages over a period of 3–4 days
  - new lesions continue to appear
  - lesions are more profuse on the head and

trunk, sparse on distal limbs and rare on palms and soles; areas of irritation (sunburn, napkin dermatitis) are often heavily involved
- tends to be more profuse in adults and in persons with cellular immune deficiency
- crusts separate in about a week. Permanent scars are rare unless secondarily infected
- *Mucosal ulcers:* not uncommon in mouth, pharynx or vagina
- *Infectious period:* the patient is infectious from 1 to 2 days before the appearance of rash to 5 days after. In immunocompromised patients, the skin lesions may remain infectious for a longer period.

Complications
- Secondary skin sepsis: due to *Streptococcus pyogenes*, less commonly *Staphylococcus aureus* — commonest complication
- Pneumonia:
  - commoner in adults (in up to 20%), particularly smokers and pregnant women
  - begins with cough and shortness of breath on day 3–5. Cyanosis, haemoptysis and in severe cases respiratory failure may develop due to extensive bilateral alveolitis. Radiologically there are discrete opacities scattered throughout both lungs, some of which may calcify after recovery (Fig. 8.1)
- Postinfectious cerebellar encephalitis (1/6000 cases):
  - often presenting only as ataxia 2–3 weeks from rash
  - complete recovery is the norm
- More extensive encephalitis, transverse myelitis and Guillain–Barré syndrome may occur rarely
- Individuals with cellular immune deficiency often have a severe illness, with numerous lesions that last longer and may become haemorrhagic. Complicating pneumonia or encephalitis are commoner
- Chickenpox in pregnancy and the risk to the newborn:
  - during the first 20 weeks: 1–2% of neonates may have low birth weight, short limbs, microcephaly, cataracts and a zoster-like rash (congenital varicella syndrome)

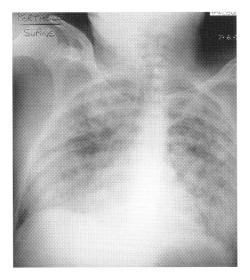

Fig. 8.1 Primary varicella pneumonitis.

- in the 2nd and 3rd trimester: infant may have active herpes zoster but no other abnormality
- a week before to a week after delivery: infant may develop severe potentially fatal chickenpox.

Diagnosis
- Doubtful cases can be diagnosed by
  - serology (fourfold rise of antibody)
  - virus culture from vesicle fluid
  - demonstration of viral antigen in vesicle scrape by immunoflorescence or PCR
- Tzanc smear (Giemsa or Wright staining of scraping from a vesicle base) may reveal multinucleated giant cells and is a rapid simple diagnostic method but of low sensitivity and does not distinguish from HSV infection
- Severe chickenpox may mimic smallpox as the eruption may be more uniform and affect distal limbs as well — electron microscopy (EM) will confirm smallpox swiftly and accurately.

Treatment and prevention
- Young children require only symptom relief and maintenance of hygiene to prevent secondary bacterial infection

• Oral acyclovir shortens illness in adults and adolescents if given within 24h of rash and is recommended

• All immunocompromised patients and patients with pneumonia should receive intravenous (IV) acyclovir.

Prevention

• Children should remain off school for 5 days from rash onset. In hospitals high-risk staff and patients should be protected from contact with chickenpox or zoster

• Zoster immunoglobulin often modifies illness if given within 10 days after exposure to chickenpox or zoster, and is recommended for:
  • antibody-negative   immunosuppressed patients and pregnant women
  • neonates whose mothers develop chickenpox in the period 7 days before to 28 days after delivery
  • antibody-negative infants exposed to chickenpox or zoster in the first 28 days of life

• Varicella vaccine: an attenuated live vaccine gives 85% protection and is safe (causing mainly mild local soreness and <5% sparse short-lived chickenpox rash within 28 days and very rare mild chickenpox in contacts) and may be used:
  • selectively (as in UK) to protect individuals vulnerable to severe chickenpox, such as
    • VZV-susceptible individuals due to receive anticancer immunosuppressive therapy or organ transplantation
    • household contacts, specifically children, of susceptible immunodeficient persons
    • HIV-infected children who are not immunosuppressed
  • universally to cover all children as part of a national immunization programme (USA, Canada, Japan and several other industrial states)

• Immunity is long-lasting. Limited observation suggests that zoster incidence is reduced

• Data are lacking as to effectiveness of acyclovir in preventing chickenpox.

Prognosis

• Childhood chickenpox is benign; occasional fatalities are from septic complications or rare encephalitis

• Most adult deaths are from pneumonia

• Case fatality rate can be 15% in immunocompromised individuals and up to 30% in severe neonatal chickenpox if not treated appropriately.

Herpes zoster (shingles)

Epidemiology and pathogenesis

• Occurs worldwide; uncommon in persons <40 years old but the rate rises progressively in >50-year-olds to an annual incidence of 10% in >80-year-olds. Commoner in patients with depressed cellular immunity (CI)

• Results from reactivation of latent VZV virus in sensory nerve ganglions. It travels down the sensory nerve to the area of skin it supplies and produces vesicles in the same way as in chickenpox

• Depressed CI from old age, disease or drugs promotes reactivation. HZ in otherwise healthy young children is probably related to less efficient development of CI at the time of the primary VZV infection, either *in utero* or postnatally.

Clinical features

• *Prodromal pain:* 2–3 days' duration on average, but may be longer

• *Rash:*
  • vesicles with surrounding erythema progress to pustules which scab and then separate
  • new lesions develop for 3–5 days
  • scabbing and separation are completed within 2–4 weeks

• *Dermatome* involved:
  • usually single
  • dorsolumbar dermatomes are most frequently involved (50%), followed by ophthalmic trigeminal, then cervical and sacral. Extremities are least commonly affected

• *Acute-phase pain:* common during rash evolution.

Complications

• Secondary skin sepsis, usually from *Streptococcus pyogenes* or *Staphylococcus aureus*

• Ocular: conjunctivitis, keratitis, uveitis, retinal necrosis, scarring of lids
• Cutaneous dissemination in the immuno-compromised. Transplant and lymphoma patients are at greatest risk (up to 40%). Visceral dissemination occurs in 5–10% of these
• Paralytic zoster, due to involvement of motor nerves, e.g. Ramsay Hunt syndrome (painful eruption in and around the ear, ipsilateral VII nerve palsy with or without vestibular disturbances), external ophthalmoplegia, bladder disturbances and weakness of limb muscles
• CNS complications: asymptomatic CSF lymphocytic pleocytosis with slightly raised protein and normal glucose occurs commonly. Meningoencephalitis, myelitis and contralateral hemiplegia due to granulomatous angitis are rare
• Postherpetic neuralgia (PHN): commonest complication, defined as persisting dermatomal pain after healing. Overall incidence is 9–15%, reaching 50% in > 60-year-olds. Pain usually subsides within 3–6 months but in a few, distressing pain persists beyond 6 months. PHN is variable in severity, type and quality.

Treatment and prevention
• Adequate analgesia is important in patients with pain
• Acyclovir, famiclovir or valaciclovir given within 72 h shorten the duration of acute illness, prevent dissemination, and probably shorten the duration, incidence and severity of PHN. The patients most likely to benefit are:
  • patients > 50 years old (more likely to develop severe acute disease and PHN)
  • patients with ophthalmic zoster, HZ of limbs and perineum (risk of paralytic zoster)
  • all immunocompromised patients with active disease at any stage (IV acyclovir)
• In established PHN, amitriptylene, topical cooling spray and transcutaneous nerve stimulation are often helpful
• Acyclovir prophylaxis helps reduce incidence of HZ following allogenic bone marrow transplant
• Trials of killed varicella vaccine are showing promise in reducing HZ incidence in transplant patients.

Smallpox
Though it has been extinct since 1977, a description of this deadly disease is important because of its possible re-emergence through a bioterrorist attack.

Epidemiology
• Smallpox virus, a single-serotype, brick-shaped DNA virus, belongs to the family of Poxviridae, only affects humans and is antigenically and morphologically similar to vaccinia, cowpox and monkeypox viruses
• The last case in the world occurred in Somalia in 1977 and global eradication was certified 2 years later
• Only the USA and Russia hold stocks of the virus in restricted laboratories. However, the possibility of other stocks in rogue hands cannot be discounted, raising the fear that the virus may be released as a biological weapon in a terrorist attack
• Transmission is by contact with respiratory secretions and skin lesions, direct, airborne or via fomite
• The IP is usually 10–12 days
• A person is infectious throughout illness and until all scabs have disappeared. Scabs remain infectious for a long time
• Mortality is high, up to 40%.

Diagnosis — clinical and laboratory
• A high index of suspicion is necessary when dealing with any case of papulovesicular eruption, if the rash is more prominent distally and there are multiple cases (this is likely in a bioterrorist attack because of point source exposure)
• Smallpox eruption appears as papules more densely distributed on the face and distal limbs (including palms and soles) after a period of fever of 3–4 days
• No more new lesions appear, papules maturing simultaneously to umbilicated vesicles, then pustules, then crusts
• A fulminant haemorrhagic form may present during the prodromal period with internal and external bleeding
• Chickenpox is the disease most likely to be confused with smallpox

- EM of vesicle fluid will rapidly diagnose smallpox (should be done in a designated laboratory).

Management
- The UK, USA and other countries have prepared strategies to meet the threat of a smallpox bioterrorist attack:
  - vaccination of key military and health-care personnel
  - contingency plans for limited ring vaccinations in the event of a localized attack
  - mass vaccination in the event of a spreading epidemic
    - vaccines are based on vaccinia virus and give immunity for many years
    - 15/1000 recipients will develop reactions: accidental inoculation vaccinia in other sites, generalized vaccinia; rarely, progressive vaccinia, encephalomyelitis; 2/1000 will die
- Quarantine of all proven and suspected cases in secure isolation facilities and their care by fully immune personnel.

Warts, molluscum contagiosum, orf and paravaccinia
Warts (verruca, papilloma)
Epidemiology and pathogenesis
- Common worldwide, warts are caused by human papillomaviruses (HPVs) of the papovavirus group (DNA viruses)
- More than 50 types are recognized. Types 1 and 2 are often associated with common and plantar warts, 6 and 11 with genital and laryngeal warts, and 16 and 18 with genital dysplasia and cancers
- Around half a million women develop cervical cancer globally each year
- Transmission is through inoculation of the skin or mucous membrane (squamous) by direct contact with a patient or autoinoculation, or indirectly from contaminated floors. Genital wart is usually sexually transmitted. Laryngeal wart is probably acquired from the mother during birth in children and orogenital sex in adults
- The IP is 3–4 months (range 1 month–2 years)

- Multiplying virus within basal cells leads to proliferation, acanthosis and keratosis. Integration of viral DNA may lead to dysplasia and neoplasm
- Immunity is probably local and cellular, leading to spontaneous disappearance. Antibodies also appear

Clinical features and complications
- Warts occur commonly on hands and feet
- In the immunocompetent they are usually harmless and self-resolve within months
- In moist genital areas, warts are often larger, fleshy and cauliflower-like (condylomata acuminata), and may involve the vaginal and anal mucosa
- Warts are often florid and protracted in cellular immunodeficiency
- Laryngeal warts in children present with hoarseness or stridor
- Cervical cancer and genital squamous cell carcinomas may develop.

Diagnosis, treatment and prevention
- Clinical diagnosis is usually sufficient but biopsy is necessary in atypical cases
- Treatment generally not necessary, unless for social stigma or painful lesions on feet or near nails
- Local salicylic acid, podophylotoxin, dinitrochlorobenzene, cryotherapy and laser are useful
- Imiquimod, an immune response modifier, also hastens recovery
- Surgery may be necessary for large genital growths
- Work on a vaccine is progressing.

Molluscum contagiosum
Epidemiology
- This is a common infection worldwide affecting children and adults. HIV patients have a high incidence
- The virus belongs to the Poxviridae family and affects only humans
- Transmission is through direct contact, sexual and non-sexual.

Clinical features and management
- Lesions are usually multiple (may be numerous in the immunocompromised). The anorectal region, trunk and face are common sites. The individual lesions are discrete, pearly papules with central umbilication
- The clinical appearance is usually diagnostic
- Disseminated cryptococcal lesions in HIV patients may appear similar
- Light microscopic examination of the cheesy material expressed from a lesion will reveal characteristic molluscum bodies (intracytoplasmic inclusions). EM will reveal the poxvirus, but is rarely needed
- Spontaneous resolution is usual. Curettage and cryotherapy are helpful in persistent cases.

Orf (contagious pustular dermatitis) and paravaccinia (milker's nodule)

Epidemiology, clinical features and management
- These are occupational diseases of humans caused by viruses belonging to the Poxviridae family which, in nature, infect sheep or goats (causing vesicular stomatitis) and cattle (causing lesions on udders: paravaccinia)
- Transmission is through contact with mucosal secretion of an infected animal
- In orf there is usually a solitary papule on a finger or other exposed areas of skin. This slowly enlarges, forming a multiloculated blister with little pain and slight surrounding erythema. Paravaccinia lesions are usually multiple, bluish papules without pustulation
- The diagnosis can be confirmed by EM demonstration of the virus in material from lesions
- Spontaneous resolution is the norm.

Streptococcal skin infections

Epidemiology
- The Gram-positive group A streptococci (*S. pyogenes*) are a common cause of skin infections worldwide. Humans are the only host
- Serotypes affecting skin are usually different from those causing throat infection
- Certain clones of M-serotypes (1 and 3) are more virulent than others and can cause invasive diseases like bacteraemia, streptococcal toxic shock syndrome (TSS) and necrotizing fasciitis
- Scarlet fever and erysipelas are commoner in temperate climates, impetigo and pyodermia in the tropics, cellulitis everywhere. In recent years invasive *S. pyogenes* diseases have become commoner in many industrialized countries
- Infection may be sporadic, endemic or epidemic
- Transmission is via direct or indirect contact with a case or carrier (nasopharynx, skin, anogenital). Preceding skin colonization is common in impetigo. Institutional outbreaks of wound infection is often nosocomial
- The IP is 1–3 days.

Pathogenesis
A range of toxins and other cellular factors are related to disease manifestations: scarlet fever rash (pyrogenic toxin A, B and C); spread of cellulitis and erysipelas (DNAse, hyaluronidase, streptokinase); invasiveness (pyrogenic toxins, proteases).

Clinical features and diagnosis
Scarlet fever
- Affects mainly young
- Rash: after a prodromal period (1–2 days) of fever and sore throat, a punctate erythematous rash appears on neck and chest, then rest of the body
- Cheeks are flushed, with pallor around mouth (circumoral pallor)
- Prominent red papillae protrude through white coating on tongue (strawberry tongue)
- Presence of follicular tonsillitis, less commonly streptococcal disease elsewhere
- Desquamation: a week later, peeling of skin on hands and feet, finer on trunk
- Diagnosis:
  - the typical rash in association with follicular tonsillitis is diagnostic
  - erythematous rash and desquamation are also seen in Kawasaki disease and TSS
  - antistreptolysin O (ASO) titre rises later

- staphylococcal scarlet fever appears to be an abortive form of TSS
- Although recurrent streptococcal tonsillopharyngitis is common, second attacks of scarlet fever are rare (toxin immunity).

Erysipelas
- Commoner in > 40-year-olds. The face (malar areas) and lower extremities are favourite sites. Portal of entry is often invisible
- A brightly red, slightly raised, tender area of skin with an clear-cut advancing edge is characteristic. Blisters may develop 2–3 days later
- On the face the rash is often on both sides of the nose, allowing differentiation from maxillary HZ
- Fever and chills are common
- Recurrence at the same site is not uncommon
- Neutrophilic leucocytosis is usual and ASO titre rises. Blister may facilitate isolation of the organism.

Cellulitis
- Results when skin and subcutaneous tissue are infected by S. pyogenes through cracks in the skin
- A rapidly enlarging area of raised tender erythema without a clear-cut margin but with projections along lymphatics is characteristic. Portal of entry may not be apparent
- May be recurrent in patients with chronic venous or lymphatic stasis
- Less commonly caused by S. aureus (a more localized process around an infected lesion, lymphangitis absent)
- Unusual bacteria:
  - following dog and cat bites—often mixed: Pasteurella multocida, Streptococcus intermedius, Capnocytophaga canimorsus (DF-2, dogs only), S. viridans, anaerobes
  - human bites—Streptococcus viridans, staphylococcal species, Eikenella corrodens, Haemophilus, anaerobes
  - anaerobic cellulitis—Clostridium welchii or mixture of anaerobic bacteria (e.g. bacterioides and Gram-negative bacilli)
  - erysipeloid—caused by Erysipelothrix rhusiopathiae, found in fish and pigs, usually an oc-cupational illness in fish handlers and slaughterers: a painful red swelling with oedema, often on hands.

Impetigo and pyoderma
- These are superficial infections of the skin usually caused by S. pyogenes, less commonly by S. aureus
- Commoner in the tropics where children living in unhygienic conditions are prone to skin colonization; minor cuts and bruises, and insect bites allow inoculation
- The face, particularly around the mouth and nose, and legs are commonly affected
- Initially a red papule appears, becoming vesicular, then pustular. Lesions break easily, exuding pus, and coalesce to form thick golden-yellow crusts. In wound and burn infection excessive redness forms around the wound with excessive discharge, which later forms a crust under which there are pockets of pus (pyoderma)
- Diagnosis is usually clinical. S. aureus should be suspected when tense pus-filled bullae are formed (bullous impetigo—usually in neonates or young children)
- Poststreptococcal glomerulonephritis may follow.

Streptococcal toxic shock syndrome
- This is diagnosed when fever, hypotension, respiratory distress syndrome, renal and hepatic impairment are associated with a documented S. pyogenes infection
- A generalized erythematous rash may be present which may desquamate
- Intravascular coagulopathy often develops
- Usually associated with necrotizing fasciitis, myositis or cellulitis, less commonly with involvement of other deep tissues
- Bacteraemia is usually present
- Motality is high (30–70%)
- Needs distinction from staphylococcal TSS (see p. 108).

Streptococcal necrotizing fasciitis
- Results from S. pyogenes infection of fascial coverings around muscles of limbs or trunk

- Site of entry can be surgical or other wounds which may be trivial
- Acute onset with fever, chills and pain at the site of infection is usual
- Locally there is marked erythema and severe tenderness spreading rapidly to the surrounding area which assumes a dusky swollen appearance
- Patient looks very ill and toxic. Hypotension, multiorgan failure and disseminated intravascular coagulopathy (DIC) are frequent
- Surgical exploration reveals fascial involvement around the muscles with inflammation and necrosis extending beyond the clinically apparent area of involvement. Muscles are usually unaffected
- Necrotizing fasciitis can also be of polymicrobial aetiology, involving a mixture of anaerobic bacteria (e.g. *Bacteroides fragilis*, anaerobic streptococci) and Gram-negative bacilli. Most such cases are associated with bowel surgery
- Isolation of *S. pyogenes* from the affected site allows distinction from non-streptococcal necrotizing fasciitis
- Blood culture is usually positive in streptococcal cases
- Mortality is high (15–20%).

Treatment

- Impetigo—topical fucidin or muciprocin in mild cases otherwise erythromycin or flucloxacillin orally (IV flucloxacillin in bullous impetigo)
- Erysipelas, cellulitis, scarlet fever—penicillin or erythromycin orally or IV; co-amoxiclav following animal/human bite
- Streptococcal TSS—IV penicillin (flucloxacillin if streptococcal aetiology uncertain), organ and haemodynamic support
- Necrotizing fasciitis—surgical debridement, IV penicillin (cefotaxime, ampicillin and metronidazole if streptococcal aetiology uncertain).

Staphylococcal skin infections
Epidemiology

- Disease-producing strains are generally coagulase-positive *S. aureus* (Gram-positive cocci)
- Humans are chief reservoir and adults are commonly carriers (15–40%)
- Anterior nasopharynx, axillae, perineum and vagina are common carriage sites
- Epidemics are often caused by specific strains, some are more virulent than others
- Transmission is by contact with a case or carrier, often via contaminated hands; indirect route rare.

Pathogenesis

- Entry into skin usually follows either a break or a plugged hair follicle
- Multiplication leads to inflammation and pus formation
- A range of toxins and cellular factors help establishment of infection, e.g. coagulase, hyaluronidase, lipase
- Other toxins are involved in disease manifestations:
  - exfoliative toxins (ETs)—scalded skin syndrome
  - toxic shock syndrome toxin (TSST-1), enterotoxin B—toxic shock syndrome.

Clinical types
Folliculitis, boils (furunculosis), carbuncle

- Boils result from infection of a hair follicle presenting as a firm tender nodule with a central pustule (folliculitis)
- Abscess results from infection spreading to adjoining areas. Buttocks, face, neck are common sites
- Carbuncles are larger abscesses with multiple heads
- Sweat glands (typically in axillae) may be infected instead of hair follicle (hydradenitis suppurativa)
- Occur worldwide; commoner in warmer climates, and in people with poor hygiene, diabetes or malnutrition
- Outbreaks of recurring infections can occur among family members and in institutions
- *Pseudomonas aeruginosa* in insufficiently chlorinated water can also cause folliculitis (hot-tub folliculitis), and waterslides, swimming pools, whirlpools, jacuzzi and other recreational facilities have been involved.

Staphylococcal scalded skin syndrome (Ritter's disease)
- Caused by ET-producing strains, usually phage type II. The toxins split the epidermis
- Rare in children > 5 years.
- Starts with an erythematous rash around eyes or mouth, spreading to trunk, then limbs
- Fever and irritability are common
- Very soon the affected skin becomes wrinkled and sloughs away, leaving a red denuded area
- Gentle pressure on an affected area produces separation (Nikolsky's sign); may be positive in apparently unaffected adjacent areas
- Large fluid-filled bullae are common
- Extensive areas may be involved, with significant loss of fluid and electrolytes and risk of death (3%)
- Lesions dry within a few days. Recovery is complete in 10 days or so
- Diagnosis is clinical, with laboratory confirmation by isolation of *S. aureus* from skin and nasopharynx
- A similar clinical syndrome in older children and adults is usually due to toxic epidermal necrolysis (TEN), caused by drug hypersensitivity (e.g. sulphonamides, barbiturates, pyrazolone derivatives). The histology is distinctive, showing splitting at dermoepidermal level.

Staphylococcal toxic shock syndrome (TSS)
- This is a multiorgan failure syndrome, caused by certain toxin-producing strains of *S. aureus*
- It has been associated with the use of tampons and intravaginal contraceptive devices; various surgical procedures, (e.g. nasal and postpartum); and skin and soft tissue infections
- Incidence of menstrual cases has declined steadily with the withdrawal of superabsorbent tampons, and currently non-menstrual cases exceed menstrual ones
- > 90% of menstrual and 50% of non-menstrual cases are related to TSST-1, others to enterotoxins (usually enterotoxin C)
- TSS is thought to be a superantigen disease. Superantigens are a class of proteins that can bypass the usual antigen-mediated immune response where antigens are intracellularly processed before presenting to T cells. Instead, superantigens bind directly to class II major histocompatibility complex and trigger overwhelming T-cell activation (5–30% of total T-cell population instead of 0.01–0.1% in the case of the conventional antigen). In TSS staphylococcal toxins act as superantigens
- Massive release of cytokines (tumour necrosis factor alpha (TNF-α), interleukin-1 and interleukin-6) follows, causing widespread capillary leak in many organ systems, resulting in their dysfunction.

Clinical features
- Acute-onset fever, watery diarrhoea, vomiting and myalgia followed by confusion, lethargy and hypotension
- Diffuse sunburn-like erythematous rash and hyperaemic mucosa (mouth, eyes, vagina)
- Evidence of two or more system involvement—muscular, hepatic, haematological, renal and central nervous systems—is commonly found
- Desquamation of skin, particularly palms and soles, 1–2 weeks later
- Complications: adult respiratory distress syndrome, acute renal failure, DIC, cardiomyopathy, encephalopathy
- Prognosis: 5% of all cases are fatal.

Diagnosis
- Clinical
- Isolation of *S. aureus* from potential sites of infection and identification of toxin genes by PCR. Blood culture rarely positive
- Differential diagnoses: scarlet fever, streptococcal TSS, Kawasaki disease.

Treatment
- Spontaneous or surgical drainage cures most boils. For carbuncles or complicated boils—flucloxacillin
- Staphylococcal scalded skin syndrome—IV flucloxacillin
- TSS—IV flucloxacillin, eradication of focus (removal of intravaginal devices or tampons, drainage of pus), organ support.

Anthrax
Epidemiology
• The aerobic Gram-positive bacillus *Bacillus anthracis* causes septicaemic disease in wild and domestic animals in many areas of Asia and Africa. In unfavourable conditions the bacilli form spores easily which can withstand drying and disinfection, and survive in animal hides and the environment for years
• Humans may become infected and develop clinical disease depending on the nature of exposure:
  • cutaneous: skin contact with infected hair, wool, hide, bones or tusks
  • inhalational: inhalation of spores aerosolized during wool processing (wool sorter's disease) or by terrorists
  • gastrointestinal: consumption of contaminated meat products
• Human anthrax is endemic in countries where animal anthrax is common
• Zoonotic anthrax is rare in most developed countries because of strict sterilization and disinfection of imported animal products; the occasional cases are from industrial or laboratory exposure or are travel-related
• However, since October 2001, 23 cases of bioterrorism-associated anthrax have occurred in the USA
• Weapon-grade anthrax spores used in bioterrorist attacks are usually highly purified and easily aerosolized, giving a larger infecting dose and more severe disease
• The IP is 2–5 days (range 1–60 days).

Pathogenesis
• After inoculation the spores rapidly germinate into vegetative forms which multiply and produce three toxins — protective antigen, oedema factor and lethal factor — leading to binding to host cells, severe oedema production and killing of host cells, respectively
• Necrosis, haemorrhage and gelatinous oedema are pathological findings. Leucocyte response is scant
• Haemorrhagic lymphadenitis may develop rapidly within mediastinum or abdomen
• Overwhelming bacteraemia or haemorrhagic leptomeningitis may develop rapidly, particularly in inhalational anthrax.

Clinical features
• *Cutaneous anthrax* usually starts as a small red papule which grows in a few days into a thick-walled, off-white vesicle with a dark red or blackish base. Later the vesicle ruptures and a black crust (eschar) forms which lasts for several weeks. Moderate to severe surrounding oedema with smaller satellite pustules are common. Pain is rare
• Bioterrorism cutaneous anthrax behaves similarly
• *Inhalational anthrax* usually presents acutely with fever and rapidly worsening stridor, dyspnoea, hypoxia and hypotension. Chest pain and pleural effusion are common. *Haemorrhagic meningitis* may be associated. Bioterrorism inhalational anthrax is often more fulminant because of the high infective dose from weapon-grade organisms
• *Gastrointestinal anthrax* presents with fever, abdominal pain, vomiting and bloody diarrhoea. Ascitis may develop rapidly
• *Septicaemic meningitis* with rapidly progressive shock and death

Diagnosis
• Cutaneous anthrax is diagnosed by demonstrating anthrax bacilli in lesion material by Gram stain, immunohistochemical stain, fluorescent antibody stain or culture, or serology
• Inhalational anthrax:
  • initially presumptive
  • differentiation is from other forms of severe respiratory distress — pulmonary embolus, viral and atypical pneumonia syndromes, plague and tularaemia pneumonia
  • massive mediastinal widening and pleural effusion are characteristic of inhalational anthrax, particularly in bioterrorism-related cases but also occur in tularaemia
  • sputum stain or culture are negative but culture of blood, urine or CSF is usually positive
  • anthrax DNA may be detected by PCR in blood, CSF or pleural fluid.

Treatment

IV penicillin or ciprofloxacin are the drugs of choice. Tetracycline and chloramphenicol can also be used.

Prognosis

Mortality is negligible in adequately treated cutaneous anthrax (10–20% if untreated); high (approaching 100%) in other forms.

Prevention

• Efficient disinfection and sterilization of all imported animal products
• Ventilation, protective clothing, maintenance of hygiene and medical supervision in high-risk premises
• Immunization of high-risk persons: an effective and reasonably safe killed anthrax vaccine is available
• Immunization of military personnel may be considered in countries deemed at risk from bioterrorism anthrax
• Environmental and personnel sampling may be necessary for investigation of bioterrorism anthrax
• Ciprofloxacin prophylaxis may be initiated when aerosolized anthrax exposure is suspected
• In endemic countries animals dying of anthrax should be buried deep in lime and people should not eat meat of ill or dead animals.

Plague

Known since ancient times as a cause of great pandemics with enormous population losses, plague is nowadays rare in nature but its potential for use as a biological weapon has brought the disease into prominence again in recent times.

Epidemiology and pathogenesis

• In nature, the causative bacterium, Yersinia pestis, is spread by infected fleas feeding on rodents (e.g. rats, squirrels, chipmunks, prairie dogs) — the natural reservoir of infection
• Foci of wild rodent plague exist in the USA, South America, Africa and south-east Asia
• 1000–3000 cases occur each year globally, 10–20 cases in the USA, none in Europe or Australia; all cases are in rural settings
• Given the availability of the organisms and the reports of development of mass production and dissemination, a bioterrorism attack of plague is a real threat
• Transmission is through bites of an infected flea but may be from another infected human with pneumonic plague via airborne transmission of aerosolized secretion
• A bioterrorist attack would most probably occur via aerosolized dissemination of weaponized Y. pestis and the resulting outbreak would be almost entirely a pneumonic one
• After cutaneous inoculation, regional lymph nodes become inflamed and swollen and may suppurate (bubonic plague). Bloodstream invasion may follow, causing septicaemia and distant localization in lungs (pneumonic plague), meninges, etc.
• The IP is usually 1–6 days (1–3 days in primary pneumonia).

Clinical features

• Fever, headache, malaise, followed by appearance of swollen tender glands
• If untreated, rapidly progresses to septicaemia, and pneumonia or meningitis
• When transmission is airborne, overwhelming pneumonia with high fever and chills, cough and bloody sputum develops rapidly.

Diagnosis, treatment and prognosis

• A high index of suspicion is necessary, particularly where there is a history of rodent exposure or when dealing with unusual clusters of tender lymphadenopathies or severe pneumonia
• Suspected patients should be isolated
• Blood cultures and lymph node aspirates should be taken for microbiology, and antibiotic treatment (streptomycin or gentamicin are drugs of choice) should begin
• Antibiotic prophylaxis for face-to-face or household contacts
• Antigen-based tests facilitate rapid diagnosis
• About 14% of USA cases are fatal; 50% of cases of pneumonic plague die.

Prevention

A vaccine is available and should be considered for laboratory personnel and persons working in potentially plague-infected areas.

Tularaemia

Caused by *Franciscella tularensis*, the disease occurs naturally as a zoonosis, affecting many small mammals in the Northern Hemisphere, and is maintained by arthropod–mammal cycles.

Epidemiology

• Type A occurs in North America and is highly pathogenic to humans, whereas the far less pathogenic Type B occurs in northern Europe and Asia
• Human infections in nature result from contact with blood or body fluids of infected animals, through arthropod bites, or through contaminated food or water
• A bioterrorist attack will, almost invariably, use aerosolized dissemination
• The IP is usually 3–5 days.

Clinical features, diagnosis, treatment, prognosis and prevention

• Natural infection most often presents as an indolent ulcer associated with swollen regional lymph glands
• Infection through inhalation often causes a primary pneumonic or a typhoidal (septicaemic) illness
• Diagnosis is by serology or rapid antigen detection in tissue samples
• Streptomycin or gentamicin are the usual drugs of choice
• Case fatality is 5–10%, chiefly from pneumonia or septicaemia
• Person-to-person transmission does not occur, so isolation is not necessary
• Antibiotic prophylaxis should be given to exposed persons
• A vaccine is not available.

Gas gangrene (clostridial myonecrosis)
Epidemiology and pathogenesis

• Gas gangrene is caused by the spore-forming anaerobic Gram-positive bacilli belonging to the *Clostridia* genus
• Most (80%) are due to *C. perfringens* (*welchii*), the rest to *C. novyii*, *C. septicum* and *C. histolyticum* which commonly inhabit human and animal intestines and are ubiquitous in soil
• Infection results from wound contamination with faeces or soil
• Occurs worldwide, mostly following street or industrial accidents, or surgery, especially above-knee amputation
• Though wound contamination is common, severe disease is rare as tissue necrosis and anaerobic conditions are necessary for growth and toxigenesis
• Of the wide range of toxins produced, α-toxin is probably the most important for progression to gangrene. It also causes haemolysis. Gas formation is characteristic
• The IP is 1–6 days.

Clinical features

• The area around the wound becomes swollen and painful; the skin colour changes to magenta and then black. Blisters are common
• Gas formation is indicated by crepitation (radiology confirms this)
• Exposed muscles look grey and the wound has an offensive, serosanguinous discharge
• The gangrene may extend rapidly and the patient becomes severely ill, with tachycardia, fever and hypotension. Renal failure and haemolytic anaemia are complications.

Diagnosis

• Presence of soft tissue gangrene, gas formation and evidence of muscle necrosis (visual, biopsy or CT) are prerequisites for diagnosis
• A Gram stain of the discharge showing clostridial organisms will be supportive but, on its own, may just indicate wound contamination or cellulitis
• Blood culture may be positive.

Differential diagnoses

As well as clostridial myonecrosis, a number of other infections can cause necrosis with

or without gas formation in and around a wound.
• Crepitant and anaerobic cellulitis: caused by clostridia or by a mixture of anaerobic bacteria such as bacteroides and anaerobic streptococci and Gram-negative bacilli
• Streptococcal and non-streptococcal necrotizing fasciitis (see p. 106).

Treatment
• Exploration and debridement of the gangrenous area or amputation of limb if spreading infection
• IV penicillin (metronidazole, if allergic) and clindamycin combination is preferred; broader cover if mixed infections suspected
• Efficacy of hyperbaric oxygen therapy is debatable
• Anti-gas gangrene horse serum has doubtful value, may cause serious reactions, and is not recommended.

Prevention
• Wounds should be thoroughly cleaned, foreign bodies removed and ischaemia avoided
• Penicillin or metronidazole prophylaxis in above-knee amputation or major trauma.

Leprosy (Hansen's disease)
Epidemiology
• This is caused by *Mycobacterium leprae* (acid-fast bacilli). Only humans are significantly affected in nature
• Following the launch of the World Health Organization's leprosy elimination programme in 1991 the prevalence has diminished considerably but the disease remains endemic in many parts of Asia, Africa and South America
• In 2000, 719 330 new cases were diagnosed; >80% live in India, Brazil, Myanmar, Madagascar and Nepal
• Infection is acquired through transmission of *M. leprae* in nasal and skin ulcer discharges of lepromatous patients via respiratory mucosa or broken skin
• There is a high degree of natural resistance to the disease and 90% of exposed people show no sign of infection.

Pathogenesis
• Initial inflammatory response at the site of inoculation is non-specific, manifesting as indeterminate leprosy
• Subsequent manifestations depend on the degree of specific CI response
• Vigorous response leads to non-caseating epitheloid granulomata involving skin and nerves, with few demonstrable organisms (paucibacillary), seen typically in tuberculoid leprosy
• Total absence of CI response leads to diffuse infiltration of skin, nerves and upper respiratory mucosa by large foam cells and numerous bacilli are present (multibacillary), characteristic of lepromatous leprosy
• Response may be anywhere between these two poles with histological picture varying from epitheloid cell to macrophage prominence. Clinical picture may be nearer tuberculoid or lepromatous leprosy
• The altered immune response is entirely leprosy specific; response to other infections is normal.

Clinical features
Indeterminate leprosy is the earliest manifestation, seen in a minority: a small pale lesion anywhere on body, without sensory loss and resolves spontaneously. Other patients present as one of the following.

Tuberculoid leprosy
• One or a few sharply demarcated hypopigmented (erythematous in pale skin) hypoanaesthetic macules anywhere on the body, progressing to raised edge with central depression and anaesthesia
• Thickened nerves, commonly ulnar, superficial radial, common peroneal and great auricular
• Neural involvement may occur alone.

Lepromatous leprosy
• Numerous skin lesions, macular, plaque-form or nodular, with ill-defined borders — symmetrically distributed
• Common on face, ears, wrists, elbows, buttocks and knees

- Skin becomes diffusely thickened and swollen, typically face, nose and lips
- Anaesthesia of skin lesions is less common
- Nasal congestion and keratitis common
- Increasing peripheral neuritis causes diffuse sensory loss.

Borderline leprosy
- Falls between tuberculoid and lepromatous, with a mixed clinical picture
- The disease state is unstable and may shift towards either pole.

Complications
- Neuropathic ulcers, deformity of face and limbs
- Secondary amyloidosis in lepromatous patients
- Gynaecomastia, testicular scarring
- *Reversal reaction (type 1 lepra reaction)*: results from increase in CI response in borderline disease causing influx of inflammatory cells into existing lesions. Skin lesions become swollen and red; neuritic and paralytic symptoms increase. Corneal anaesthesia may occur
- *Erythema nodosum leprosum (type 2 lepra reaction)*: results from vasculitis, probably TNF-mediated, triggering neutrophilic infiltration. Occurs in lepromatous and borderline lepromatous states, usually after few months of treatment. Crops of tender subcutaneous nodules are associated with fever and arthralgia. Iridocyclitis and neuritic symptoms may develop.

Diagnosis
- Leprosy is diagnosed when a patient has one or more of the following:
  - hypopigmented or reddish skin lesion(s)
  - involvement of the peripheral nerves, as demonstrated by definite thickening with loss of sensation
  - skin smear positive for acid-fast bacilli
- Typical histology in skin or nerve biopsy is diagnostic but may be inconclusive. Immuno-histopathological methods may prove more reliable
- PCR demonstration of DNA is highly sensitive and specific but available currently only as a research tool.

Treatment
- Multibacillary leprosy (lepromatous, borderline lepromatous): a combination of rifampicin, clofazimine and dapsone for 2 years
- Paucibacillary leprosy (tuberculoid, borderline tuberculoid): rifampicin plus dapsone for 6 months
- Treatment should continue during either type of reaction plus:
  - type 1: corticosteroids
  - type 2: mild—aspirin or chloroquine; severe—corticosteroids, thalidomide (not in women of child-bearing age)
- Surgery for deformity and rehabilitation.

Prognosis
- Leprosy is a leading cause of long-term neurological disability
- It is associated with high levels of stigma and social exclusion.

Prevention
- Examine family members and other close members regularly for signs of leprosy
- Bacille Calmette–Guérin vaccination gives some protection against leprosy.

Lyme disease (Lyme borreliosis)
Epidemiology
- Caused by *Borrelia burgdorferi*, a spirochaete, the disease is focally endemic in the forests and woodlands of temperate North America, Europe and Asia where both the insect tick vector *Ixodes* and animal hosts, rodents and deer, abound in the moist leafy undergrowth
- It is the commonest vector-borne illness in the USA
- Immature ticks become infected while feeding on rodent hosts during summer; infected adults feed on deer
- Human infections mostly occur during summer through tick bites; transmission unlikely if the tick had fed for <36 h
- The IP is 7–14 days (range 3–30 days)
- Infection may remain asymptomatic.

Pathogenesis
- After inoculation the spirochaetes may initially multiply and migrate outwards, producing the typical erythema migrans
- Early dissemination occurs within days or weeks via blood or lymphatics to skin, musculoskeletal, cardiovascular or nervous systems with local lymphocytic inflammatory and sometimes vasculitic response.

Clinical features
- Erythema chronicum migrans (ECM), a red papule gradually enlarging to a large red area with some central fading, is often the first sign but absent in 25% of cases. A tick bite is often not remembered
- Malaise, headache and arthralgia are common
- Within days or weeks of ECM, and sometimes without it, patient often develops:
  - multiple ECM-like lesions accompanied by fever, fatigue, chills, headache
  - migratory myalgia and arthralgias
  - less commonly, lymphocytic meningitis, cranial nerve palsies, polyneuropathy or myelitis. In Europe and Asia, radiculitis with lymphocytic pleocytosis but without meningism is commoner
  - rarely, cardiovascular manifestations — conduction defects, myocarditis
- These early signs of disseminated infection often resolve within weeks or months, even without treatment, but may progress to:
- Late Lyme disease characterized by joint (chronic or recurrent arthritis), neurological (chronic encephalopathy, polyneuropathy) or skin (acrodermatitis chronicum atrophicum — patchy discoloration on limbs becoming atrophic in time) manifestations.

Diagnosis
- Treatment may be initiated on the basis of the clinical picture and exposure history
- ECM-like rash has also been seen recently in south/south-eastern states of the USA following bites from a tick called *Amblyomma americanum*. The illness has been named Southern tick-associated rash illness (STARI), and is probably caused by a newly identified spirochaete named *B. lonestari*. Serology for Lyme disease is negative
- Most patients have antibodies by the time of early disseminated stage but may be negative in the early stage
- Initially IgM antibodies then IgG antibodies develop which may persist for months or years after treatment
- Initially an ELISA or indirect fluorescent antibody test is performed; positive or doubtful tests are corroborated by Western blot (IgG and IgM)
- Seropositivity alone is not a marker for active infection, thus differential diagnosis of late-stage disease from other chronic joint and neurological diseases or chronic fatigue syndrome can be difficult
- The organism can often be cultured from early EM lesion but facilities are limited
- PCR detection of *B. burgdorferi* DNA in blood, CSF or synovial fluid is under research.

Treatment and prognosis
- Doxycyline for 10–30 days (or amoxicillin or third-generation cephalosporin) is usually effective; longer courses should be given for late-stage arthritis
- IV cefotaxime is preferred for neurological disease
- Lyme disease is rarely fatal but permanent joint, nervous system or cardiovascular system disabilities may occur.

Prevention
- Avoid tick-infested areas in spring and summer if possible, otherwise use personal protection
- Cover arms and legs, wear high rubber boots and apply insect repellents containing DEET (diethyltoluamide)
- Perform a daily tick check and remove any found with a tweezer. Do not use heat or other products
- Antibiotic prophylaxis is generally not necessary in the absence of symptoms
- A vaccine is not currently available (withdrawn in 2002).

Fungal skin infections
Dermatophytosis (ringworm, tinea)
Epidemiology
- Caused by various species of dermatophyte fungi, *Microsporum, Trichophyton* and *Epidermophyton*; occurs worldwide; also affects animals
- Transmission is by direct contact with skin or hair or indirectly via combs, chairs, shower floors or bath areas; sometimes from animals, rarely from soil.

Clinical features and diagnosis
- Ringworm of body (tinea corporis) and groin (tinea cruris): usually starts as a red scaling area which enlarges with central clearing. More acute lesions are itchy and vesicular
- Ringworm of scalp (tinea capitis): localized alopecia with broken hair shafts and scaling. Beards may be affected, presenting as folliculitis which may be pustular (tinea barbae)
- Nail infections (tinea unguum): dull, brittle and discoloured nails, often distorted and thickened
- Foot infection (tinea pedis, athlete's foot): scaling or fissuring of soles or toe webs. Vesicles may form
- Diagnosis is by demonstrating fungi in skin scrapings, hair or nail clippings by KOH smear and culture.

Treatment
- Topical clotrimazole, miconazole, econazole or terbinafine is usually sufficient for mild localized infections
- Oral itraconazole or terbinafine is preferred in extensive involvement and when nail or scalp are involved
- Topical amorolfine is an alternative for nail infections.

Malassezia infection (pityriasis versicolor)
- The yeast *Malassezia furfur*, a commensal of human skin, may flourish in warm humid conditions and produce well-demarcated hypo- or hyperpigmented macules on the arms and upper trunk. Lesions often coalesce to form large areas

- Fine scales may be visible and folliculitis may develop
- Clinical diagnosis is confirmed by demonstrating the organisms in KOH-treated skin scrapes
- Selenium sulfide shampoo or azole creams are effective; oral azole is reserved for florid cases
- Pigment changes may persist for months post treatment.

Mycetoma (Madura foot)
Epidemiology and pathogenesis
- Caused by actinomycetes (e.g. *Nocardia, Actinomadura, Streptomyces*) or fungi (e.g. *Madurella mycetomatis*). Generally seen in tropics or subtropics — mostly in Central and South America, Africa and India
- *Nocardia* is often the cause in Central America and *Madurella mycetomatis* in India and Africa
- The organisms live in soil (actinomyces) or plants (fungi) and enter skin through trauma, mostly of foot
- Chronic suppurative infection over many years is the rule. Within the pus, colonies of organisms clump together to form grains. Deep tissue destruction is common.

Clinical features
- Initially a painless subcutaneous lesion forms, mostly in foot
- This gradually enlarges to form a large woody painless swelling, later sinuses appear discharging pus
- Systemic symptoms and dissemination are rare.

Diagnosis and treatment
- Needs differentiation from chronic osteomyelitis
- Diagnosis requires demonstration of grain in pus or biopsy and distinction between fungal and actinomycetes causes as treatment differs
- Actinomycetes often respond to prolonged antibiotic therapy with streptomycin or rifampicin with co-trimoxazole

• Fungal causes are difficult to treat. Perhaps half would respond to azoles. Amputation may be necessary.

Cutaneous infestations
Scabies
Epidemiology and pathogenesis
• Scabies is caused by a mite *Sarcoptes scabiei*, occurs worldwide, and is commoner in resource-poor countries and in conditions of overcrowding, lack of personal hygiene and sexual promiscuity. All ages are affected
• Reservoir is humans
• Transmission is by close skin contact, typically by holding of hands or in bed; not by bedding or clothing
• The gravid adult female mite burrows underneath the stratum corneum, laying eggs which mature into adults in about 14 days
• Adults emerge onto skin to spread to other parts of the body or to infect other hosts
• The cardinal symptoms of scabies, rash and itching, result from sensitization to mite deposits in the burrow. This takes several weeks to develop, so initially scabies is asymptomatic. Inflammatory cells accumulate around the burrows to form papules or plaques and distant hypersensitivity rash may develop
• Infestation is cut short by possible cellular immunity and physical removal through scratching. People with cellular immunodeficiency (e.g. AIDS patients) and those unable to scratch efficiently because of senility or psychiatric illness may develop superinfestation with huge number of mites. They are highly infectious and can cause outbreaks.

Clinical features
• The usually papular lesions with associated scratch marks are symmetrically distributed, often on fingers, wrists and buttocks, around the waist or on genitalia. Vesicles or pustules may form or may become eczematous
• The characteristic lesion is the burrow—a thin dark line ending in a pinhead blister (which contains the mite)
• Itching is usually worse at night.

Complications
• Norwegian or crusted scabies: in superinfestation, crusted psoriasiform lesions develop widely over the body; itching and burrows are usually absent
• Impetigo from secondary infection with *S. pyogenes* is common in the tropics.

Treatment and prevention
• Topical malathion or permethrin are the preferred agents; benzyl benzoate is an irritant and less effective
• Single-dose ivermectin is useful in Norwegian scabies where local treatment may not succeed
• The whole family and sexual partners should be treated
• The itching of scabies may persist for several weeks post treatment; crotamiton application may help.

Pediculosis (louse infestations)
Epidemiology and pathogenesis
• The three species of human louse, *Pediculus humanus capitis* (head louse), *P. humanus corporis* (body louse) and *Pthirus pubis* (crab louse), occur worldwide
• Only *P. humanus corporis* transmits infection, e.g. epidemic typhus, louse-borne relapsing fever, trench fever
• They live on hair or clothing (body lice) and feed on human blood, the bite causing intensely itchy maculopapules from sensitization. Eggs (nits) are firmly attached to hair or clothing. Lice leave febrile hosts or corpses, thus encouraging transmission of infection by body lice during epidemics
• All forms are transmitted by direct contact (sexual in the case of crab lice) or by sharing of clothes (body lice)
• Head lice are common in children; school outbreaks may occur. Body lice occur mainly in people living in overcrowded, unhygienic conditions, not changing clothes, as in war or disaster conditions, and are transmitted by direct contact or sharing of clothes. Crab lice occur mainly in the sexually active.

Clinical features, diagnosis and treatment
• Head lice: pruritic excoriated lesions on scalp, neck and shoulders; secondary bacterial infections are common, resulting in regional lymphadenitis
• Body lice: lesions can be anywhere
• Crab lice: tiny bluish lesions, intensely pruritic, in the pubic area. May affect eye lashes and axillae
• Diagnosis is by finding nits or adults on hair or clothing
• Carbaryl, malathion and pyrethroids (for head lice) and carbaryl or malathion (for crab lice) are generally used
• Lotion, liquid or cream formulations should be used with a contact time of 12 h and reapplication after a week
• In body lice outbreaks, clothing and bedding should be laundered in hot water or dry cleaned, or treated with malathion or pyrethroids and washed.

Cutaneous myiasis
Epidemiology and pathogenesis
• This is caused by the maggots (larvae) of several kinds of flies
• The larvae require a period of maturation under the skin of warm-blooded animals, and humans become involved in appropriate circumstances
• The African tumbu fly (tropical Africa) and human bot fly (South and Central America) are usually involved, less commonly horse and cattle bot flies
• Eggs of the tumbu fly deposited on sandy soil or drying clothes mature quickly on contact with warm skin into larvae which penetrate into skin; human bot fly eggs are transported to warm-blooded hosts by blood-sucking insects directly from the females. Under the skin larvae mature for 6–12 weeks, then drop to the ground and pupate
• Eggs of horse and cattle flies are laid on the hair of the host or on grass, and on contact with warm skin mature into larvae and penetrate. In humans they cannot mature further but may migrate in the epidermis for weeks (horse bot fly) to produce creeping dermal myiasis or penetrate deeper to produce boil-like lesions (cattle bot fly).

Clinical features, diagnosis and treatment
• Tumbu, human and cattle bot fly myiasis present as a 'boil', with a black central spot representing the spiracle of the larva. Blocking the opening with an oily substance suffocates the larva and it can be extracted with forceps as it wriggles
• Horse fly myiasis resembles the lesion of cutaneous larva migrans (see p. 243). Larvae can be identified through smeared mineral oil and then extracted with a needle.

Tungiasis
• Seen in west and east Africa and the Indian subcontinent. *Tunga penetrans* (jigger flea, sand flea, chigoe) is the cause
• Gravid females developing from larvae living on soil penetrate into the skin of an animal (often pig or poultry) or human, usually through cracks in the soles of the feet
• The flea feeds and swells to a 'pustule' with a central black spot where its posterior spiracle is exposed. Pain and irritation develop as the flea distends further. The eggs are laid, the skin ulcerates and the flea is expelled
• Multiple pit-like ulcers on soles or around nails are common
• Treatment is removal of the flea without bursting it as this may lead to severe inflammation.

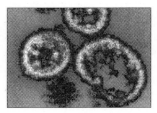

Infections of the Gastrointestinal Tract

Clinical syndromes

Acute community-acquired infective diarrhoea

Major presenting features: *acute-onset diarrhoea with or without abdominal pains and vomiting.*

Infective diarrhoeas are a major cause of morbidity and mortality in the developing countries, particularly among children. Poor standards of hygiene and sanitation, malnutrition, overcrowding and poor medical resources are the responsible factors. The disease is also very common in the developed countries, but of low mortality.

Common causes
See Table 9.1.

Distinguishing features
In clinical practice it is useful to distinguish between inflammatory and non-inflammatory diarrhoeas. This often gives a clue to the likely pathogens (Table 9.2).

NB. Diarrhoea is often non-inflammatory in the early stages when due to *Shigella*, *Salmonella* and *Campylobacter*.

Complications
- Dehydration and renal failure
- Hypernatraemic dehydration of infants:
  - high serum sodium (> 150 mmol/L) despite overall body deficit of sodium, because of greater loss of water than sodium
  - concurrent hypertonicity of intracellular compartment may cause brain damage
- Septicaemia (*Salmonella*, *Yersinia*, *Campylobacter fetus*)
- Toxic colonic dilatation (*Salmonella*, *Campylobacter*, *Shigella*, *Clostridium difficile*)
- Haemolytic–uraemic syndrome (entero-haemorrhagic *E. coli* O157, *Shigella dysenteriae*)
- Reactive arthritis (*Shigella*, *Salmonella*, *Campylobacter* particularly in HLA-B27-positive individuals)
- Erythema nodosum (*Salmonella*, *Campylobacter*, *Yersinia enterocolitica*)
- Persistent diarrhoea (see box, p. 120).

	Developing countries	Developed countries
Children	Rotavirus ETEC Adenovirus EPEC Giardia Shigella EAggEC Campylobacter Salmonella	Rotavirus Adenovirus SLVs* Campylobacter Shigella Salmonella Cryptosporidium VTEC
Adults	ETEC Salmonella Shigella NLVs Vibrios Cryptosporidium Giardia	Campylobacter Salmonella NLVs Giardia Cryptosporidium VTEC

* Sapporo-like viruses (SLVs) have emerged as an important cause of community-acquired diarrhoea in young children in industrialized countries worldwide, prevalence data from developing countries lacking. They are caliciviruses, similar to Norwalk-like viruses (NLVs).

EAggEC, enteroaggregative E. coli; EPEC, enteropathogenic E. coli; ETEC, enterotoxigenic E. coli.

Table 9.1 Common causes of acute community-acquired infective diarrhoea.

Features	Inflammatory diarrhoea	Non-inflammatory diarrhoea
Character of stools	Usually small volume, containing blood and pus (naked eye/microscopy)	Large volume, watery without pus or blood (naked eye/microscopy)
Pathology	Inflammation of colonic and/or distal ileal mucosa	Proximal small intestine
Mechanism of diarrhoea	Mucosal inflammation impairs absorption of fluid and possible secretagogue effect of inflammatory products	Secretory/osmotic diarrhoea induced by enterotoxin or other mechanisms. Mucosal inflammation absent
Likely pathogens	Shigella, Salmonella, Campylobacter, E. coli O157, EIEC, Clostridium difficile, Yersinia enterocolitica, E. histolytica	Cholera, ETEC, EPEC, toxin-type food-poisoning, rotavirus, adenovirus, NLVs, Cryptosporidia, Giardia

EIEC, enteroinvasive E. coli; EPEC, enteropathogenic E. coli; ETEC, enterotoxigenic E. coli; NLV, Norwalk-like virus.

Table 9.2 Distinguishing features of inflammatory and non-inflammatory diarrhoea.

PERSISTING DIARRHOEA AFTER AN ACUTE ONSET

• Persistence of infective diarrhoea beyond 2 weeks is rare except in parasitic infections and in tropical infantile diarrhoea due to EPEC and EAggEC
• In other cases, suspect:
  • underlying inflammatory bowel disease (if stools are bloody)
  • postinfective irritable bowel syndrome
  • postinfective intestinal malabsorption (if history of travel to the tropics)
  • secondary lactose malabsorption (in infants).

Management of acute community-acquired diarrhoea

• Most patients have mild to moderate diarrhoea of short duration and require only attention to hydration
• Arrange stool culture and faecal microscopy (to distinguish between inflammatory and non-inflammatory diarrhoea) if diarrhoea is severe, watery or bloody or patient is in high-risk group for complications (i.e. elderly, debilitated or immunocompromised) or has other concurrent illnesses
• All ill patients should have blood cultures
• Antibiotics should be avoided in average cases of bacterial diarrhoea as the duration of illness is at best only marginally shortened and antibiotics may encourage *in vivo* bacterial resistance
• Consider empirical fluoroquinolone therapy if patient has severe inflammatory diarrhoea or belongs to a high-risk group (invasive salmonellosis more likely)
• Consider floroquinolone therapy if stool culture shows *Shigella*, *Salmonella* or *Campylobacter* and patient has continuing severe diarrhoea
• Antimicrobial therapy is also indicated in the following circumstances:
  • invasive salmonellosis
  • cholera
  • severe yersiniosis
  • giardiasis
  • amoebiasis
  • traveller's diarrhoea

• Empirical antibiotic therapy is rarely indicated in infantile diarrhoea as most is of viral aetiology
• Avoid antimotility drugs (except for very short courses in adult traveller's diarrhoea).

Prevention

• In hospitals, all patients should be nursed under enteric precautions
• All cases should be excluded from work, school or day nurseries until symptom free
• Food handlers whose work involves touching unwrapped foods to be consumed raw or without further cooking, and who are excreting pathogens should have three consecutive negative stools at weekly intervals before returning to work
• Prolonged exclusion of young children excreting pathogens from day nurseries, playgroups or nursery schools is not necessary if supervision of hygiene is adequate
• Meticulous attention is necessary to standards of personal hygiene, hand-washing facilities and supervision of hygiene in childcare facilities
• Public water supplies should be protected from human and animal faeces and should be filtered properly if such risks exist
• Travellers should drink only boiled or bottled waters, avoid unfiltered lake or stream water (water-purifying tablets are unreliable against parasitic diarrhoea), eat only reliably washed salads, raw vegetables and fruits and freshly cooked foods, and avoid raw milk
• Foods of animal origin should be thoroughly cooked
• Eggs should be boiled for at least 3 min
• Cooked food should be promptly refrigerated
• High standards of hygiene must be observed in all areas where raw meat is handled, i.e. slaughterhouses, meat shops, restaurants and domestic kitchens. Raw and cooked food should be kept separate. Frozen food needs complete thawing before cooking
• Children should be educated and supervised during farm visits with regard to hygiene and hand-washing

• In developing countries, provision of safe drinking water, sanitary disposal of excreta and public education on food hygiene are important measures.

Specific infections

Helicobacter pylori infection
Epidemiology
• Is the commonest cause of peptic ulcer and a major cause of non-cardia gastric adenocarcinoma
• The spiral-shaped Gram-negative bacilli live under the mucous layer of the stomach and duodenum; humans are the main reservoir (also found in monkeys and cats)
• Found worldwide; 50% of adults >50 years are infected in the developed countries. Prevalence is higher in the developing countries
• Prevalence of infection and incidence of ulcer and gastric cancer are declining in the developed countries
• Infection is usually acquired in early childhood and persists unless treated
• Transmission is probably person to person, faecal–oral or oro-oral in early childhood through direct contact or poor hygiene.

Pathogenesis
• Infection damages the protective mucous layer in weeks or months and causes chronic superficial gastritis or duodenitis
• Exposure to acid over a long period can lead to ulcer formation or atrophy, metaplasia, then dysplasia, and eventually cancer.

Clinical manifestations
• Most infections are asymptomatic despite the presence of gastric mucositis
• Around 15% will develop peptic ulcers
• Non-cardia gastric adenocarcinoma is 2–6 times commoner in people infected with *H. pylori*
• Risks for gastric lymphoproliferative malignancies are also higher
• It is not clear whether infection can cause non-ulcer dyspepsia.

Diagnosis
• Evaluation of possible ulcer patients is carried out by endoscopy so that both ulcer and infection can be confirmed but patients may initially be screened by checking for antibodies
• At endoscopy, biopsy specimen is examined for *H. pylori* by urease test and histology, and culture
• Breath tests with tagged urea are usually used to assess effectiveness of treatment; serology is also useful.

Treatment and prevention
• One-week triple therapy (two antibiotics plus a proton pump acid inhibitor) course is effective in 90%
• Amoxicillin and clarithromycin are preferred combination for initial treatment
• Use amoxicillin and metronidazole for treatment failures
• For non-ulcer dyspepsia, benefits of treatment are unclear
• General food and personal hygiene should be followed.

Shigella infection (shigellosis, bacillary dysentery)
Epidemiology
• The Gram-negative shigella bacilli, which are strict human pathogens, have four subgroups: *S. dysenteriae, S. flexneri, S. boydii* and *S. sonnei*
• Occurs worldwide and is the third commonest cause of bacterial diarrhoea in the developed countries
• In the developed countries most cases are due to *S. sonnei*, the rest mainly due to *S. flexneri* of imported origin. Outbreaks are common in infant schools, daycare centres, prisons and camps
• In the tropics, endemic dysentery is most commonly due to *S. flexneri*, followed by *S. sonnei*
• Epidemic dysentery is due to *S. dysenteriae* type I (Shiga bacillus, Sd1) and occurs mainly in tropical countries
• Transmission is faecal–oral from cases with diarrhoea; may be water- and food-borne or sexually transmitted through oro-anal contact. Flies may spread infection in the tropics

- Infecting dose is low (10–100 organisms), allowing for easy person-to-person spread
- Children are mainly affected. Overcrowding and poor personal hygiene encourage transmission
- Incubation period (IP) is 2–4 days (range 1–7 days).

Pathogenesis
- After entry, the organisms invade the colonic mucosa causing inflammation, ulceration, haemorrhage and sloughing, and fluid secretion
- Sd1 also elaborates an enterotoxin (Shiga toxin) which can cause microangiopathy leading to haemolytic–uraemic syndrome and thrombotic thrombocytopenic purpura.

Clinical features
- Many infections are either inapparent or mild
- The illness begins with malaise, abdominal discomfort and watery diarrhoea
- In many S. sonnei infections, the diarrhoea may remain watery and settles within 3–5 days
- In others, and in most S. flexneri, S. boydii and Sd1 infections blood and mucus appear in stools within a day or two
- In severe cases the classic picture of dysentery appears: fever and severe abdominal cramps, tenesmus and passage of very small volume stools consisting just of blood, pus and mucus
- Convalescent carriage after clinical recovery is common but usually ceases by 8 weeks.

Complications
- Rare in average cases of S. sonnei infection
- Toxic dilatation and/or perforation (S. flexneri and Sd1)
- Haemolytic–uraemic syndrome (Sd1).

Diagnosis
Stool culture is required if differentiation is necessary from other causes of inflammatory diarrhoea.

Treatment
- Most cases are mild and require only attention to adequate oral fluid intake

- Severe infections require antibiotics:
  - drug resistance is common, so knowledge of local resistance pattern is useful
  - ciprofloxacin for adults and trimethoprim for children are often adequate.

Prognosis
Approximately 5–15% of epidemic Sd1 cases are fatal. Fatality rare in other forms.

Prevention
See p. 120.

Campylobacter infection
Epidemiology
- Most human infections are due to *Campylobacter jejuni* of the *Campylobacter* species (Gram-negative bacilli).
- Occurs worldwide and is the commonest cause of bacterial diarrhoea in the developed countries
- Commonest in children and adults in the developed countries and children <2 years in the developing countries
- It is a zoonotic infection and reservoirs for human infections are gastrointestinal tracts of birds (particularly poultry) and animals, cattle and pets
- Transmission is through consumption of raw or undercooked meat, unpasteurized or contaminated (bird-pecked) milk or water, or contact with infected pets
- Person-to-person transmission is rare
- The IP is 3–5 days (range 1–10 days).

Pathogenesis
- Invasion of the lower intestines, mainly ileum and colon, leads to mucosal inflammation and fluid secretion
- Bacteraemia is rare.

Clinical features
- Infection may be inapparent
- Fever, myalgia and abdominal pain may precede diarrhoea by 1–2 days
- Stools are initially large-volume and watery but later may become small-volume, bloody and mucoid

- Vomiting is common in children
- The diarrhoea ceases within 1 week, but low-grade abdominal pains may continue for several more days
- Most patients cease to excrete organism in their faeces within 3–4 weeks.

Complications
- Acute ileitis with pain and tenderness in the right lower abdomen, which may mimic acute appendicitis
- Severe colitis
- Guillain–Barré syndrome
- Rarely: acute cholecystitis, erythema nodosum and reactive arthritis.

Diagnosis
Differentiation from other types of inflammatory diarrhoea requires stool culture.

Treatment and prognosis
- Most cases require only prevention or correction of dehydration
- Antibiotics are only indicated in patients with severe or prolonged diarrhoea
- The drugs of choice are either erythromycin or ciprofloxacin for 5 days
- Fatality is rare.

Prevention
See p. 120.

Non-typhoid *Salmonella* infection (salmonellosis)
Epidemiology
- Salmonellosis is caused by the non-typhoid salmonellae (Gram-negative bacilli) which are primarily parasites of animals, inhabiting their intestines
- Of about 2000 serotypes, only a small number account for the vast majority of human infections. Common examples are S. enteritidis, S. typhimurium, S. virchow, S. hadar, S. heidelberg, S. agona and S. indiana
- It is the second commonest cause of bacterial diarrhoea in the developed countries, where the incidence has increased substantially in the last three decades

- The increase is related to large-scale intensive farming methods for food animals which create conditions suitable for the rapid spread of infection among the animals and opportunities for human infections along the food chain from farm to kitchen
- The majority of cases are due to S. enteritidis and S. typhimurium
- A high proportion of typhimurium cases are caused by S. typhimurium DT 104 which may have emerged due to the use of antibiotics in animal husbandry; illness is often more severe and invasive
- Transmission is through consumption of inadequately cooked contaminated meat (mainly poultry, beef and pork) or hens' eggs (infected via the oviduct), and raw milk
- Person-to-person transmission is infrequent. Most institutional outbreaks are food related but may follow admission of patients with undiagnosed diarrhoea, particularly in geriatric or maternity/neonatal units
- Turtles and other pets are occasional sources
- Little is known about the epidemiology in developing countries but incidence has risen in a number of countries in the Middle East and south-east Asia. Nosocomially transmitted hospital outbreaks of multidrug-resistant salmonella are not uncommon in Africa and India
- The IP is 12–48 h (range 6–72 h).

Pathogenesis
- Salmonella invades lower intestinal mucosa, mutiplying locally and producing inflammation and secretion of fluid
- Transient bacteraemia is not uncommon
- Signficant invasive disease with septicaemia or organ involvement is commoner in infancy and old age, in patients with immunosuppression, debility or achlorhydria, and with certain strains, e.g. multidrug-resistant S. typhimurium DT 104, S. cholerae-suis, S. virchow and S. dublin.

Clinical features
- Infection may remain inapparent
- The illness begins abruptly with nausea, vomiting, malaise, cramp-like abdominal pains and diarrhoea

- Initially, the stools are large-volume and watery without blood but later may become blood-stained and mucoid
- The severity of diarrhoea is variable, from a mild attack lasting no more than a few hours to frequent voluminous watery stools
- Diarrhoea usually settles within a few days; persistence beyond 3 weeks is rare
- After recovery, patients usually excrete salmonella for 4–6 weeks, but longer in extremes of ages. Chronic carriage beyond 1 year is rare (< 1%).

Complications

Severe diarrhoea, dehydration and renal failure: prone to occur in the elderly, the immunocompromised and in persons with gastric achlorhydria.

Colitis: with severe bloody diarrhoea, may be segmental, mimicking Crohn's disease (Fig. 9.1). Rarely, toxic megacolon may develop.

Ileitis: with pain and tenderness localizing over the right iliac fossa. This may be confused with appendicitis.

Invasive salmonellosis: this may present as either

- septicaemia, or
- typhoid-like illness without significant diarrhoea, or
- metastatic infection in extraintestinal sites (e.g. meninges, bones and joints, lungs, spleen, kidneys, heart valves and atheromatous blood vessels).

Postinfective irritable bowel syndrome.

Reactive arthritis: affecting one or more joints (commonly knee and ankles) during the 2nd or 3rd week of illness, particularly in individuals with HLA-B27 antigen.

Diagnosis

Differentiation from other causes of inflammatory diarrhoea requires stool culture.

Treatment and prognosis

- Most cases have a short-lasting, self-limiting illness and require only attention to hydration
- Patients with, or at high risk for, invasive illness, and those with severe symptoms, should receive antibiotics (ciprofloxacin in adults, cefotaxime or ceftriaxone in children)
- Deaths are rare except in infancy or old age due to severe or metastatic disease.

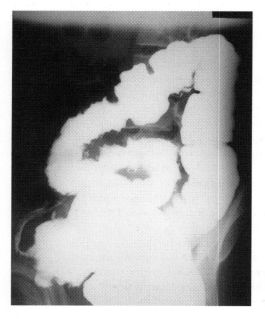

Fig. 9.1 Predominantly right-sided colonic involvement in a 17-year-old girl with severe salmonella colitis.

Prevention

See p. 120.

Verocytotoxin-producing
Escherichia coli (VTEC) infection

Epidemiology

• VTEC (also known as enterohaemorrhagic *E. coli*) strains are an important cause of diarrhoea in Europe and North America, and of diarrhoeal haemolytic–uraemic syndrome (HUS)
• VTEC O157 is the strain most commonly involved; other serotypes are occasional causes
• Incidence has increased markedly in recent years
• Most cases are sporadic (>80%)
• Incidence is highest among children <5 years, but all ages are susceptible
• It is a zoonotic infection; the main reservoir is the intestines of cattle, also sheep and other animals
• Infecting dose is very low (<100 organisms) so it is easily transmitted
• Transmission is by consumption of contaminated food (undercooked beefburgers commonly involved), unpasteurized milk and milk products or water, and contact with animals (farm visits) and infected persons; contaminated swimming pools and contaminated fruit juices are other sources
• The IP is 3–4 days (range 2–14 days).

Pathogenesis

• The verocytotoxin elaborated is similar to Shiga toxin of *S. dysenteriae* type 1
• Causes destruction of colonic cells and damage to capillary endothelium, producing haemorrhagic colitis and microvascular angiopathy, particularly affecting glomeruli and central nervous system (CNS).

Clinical features and complications

• Spectrum of illness varies from mild non-bloody diarrhoea to haemorrhagic colitis (50% of cases)
• The latter is characterized by severe abdominal pains and passing of frank bloody stools; fever is rare
• Illness is usually self-limiting and resolves

within 7 days but about a third require hospital admission
• HUS:
  • develops in 2–7% of cases overall, up to 30% in some outbreaks
  • develops between 2 and 14 days (median 6 days) after onset of diarrhoea
  • oliguria, renal failure, thrombocytopenia and microangiopathic haemolytic anaemia develop
  • seizures, coma and hemiparesis may be present
• Thombotic thrombocytopenic purpura may complicate VTEC diarrhoea, particularly in adults
• Faecal excretion of organisms often ceases after a few days but may be prolonged in young children.

Diagnosis

• In laboratories, all diarrhoeal stool samples should be examined for *E. coli* O157 irrespective of age of patient
• All isolates should be sent to reference laboratory for confirmation, phage typing, toxin typing and subtyping
• Patients with suspected acute inflammatory bowel disease, patients passing bloody stools and patients with HUS should be investigated for non-O157 VTEC if their stools are negative for *E. coli* O157
• Serodiagnosis is useful in investigating culture-negative cases of HUS and haemorrhagic colitis as stools often become negative after 4 days.

Treatment

• Attention to hydration, monitoring of renal function, blood counts and blood films are essential
• Antibiotic treatment may not be effective
• HUS and CNS complications should be appropriately managed.

Prognosis

• Around 4% cases overall are fatal but outbreaks involving the elderly have higher death rates

- 3–17% of HUS cases are fatal
- Long-term renal or neurological sequelae are not uncommon.

Prevention
See p. 120.

Diarrhoea due to other *E.coli*
Epidemiology
- Four other pathogenic types are also important causes of diarrhoea, and affect humans only:
  - enteropathogenic (enteroadherent) *E. coli* (EPEC)—a major cause of infantile diarrhoea in the developing countries; nursery outbreaks common
  - enterotoxigenic *E. coli* (ETEC)—a common cause of infantile diarrhoea in developing countries and an important cause of traveller's diarrhoea in visitors to these areas
  - enteroaggregative *E. coli* (EAggEC)—an important cause of childhood persistent diarrhoea in the developing countries
  - enteroinvasive *E. coli* (EIEC)—an uncommon cause of dysentery-like illness in the developing countries
- Transmission is faecal–oral, often through contaminated infant feeds (EPEC and ETEC), contaminated food and water (ETEC and EIEC) or contaminated hands (ETEC nursery outbreaks)
- The IP is 1–2 days, may be shorter.

Pathogenesis
EPEC: organisms adhere to small intestinal mucosal cells causing dissolution of microvilli.
ETEC: diarrhoea results from the action of two toxins: heat-labile toxin (LT) stimulates adenylate cyclase causing increase of cyclic adenosine monophosphate inside cells; heat-stable toxin (ST) stimulates guanylate cyclase causing increase of guanosine monophosphate. A net increase in intestinal secretion follows.
EAggEC: unclear.
EIEC: invades ileal and colonic mucosa causing inflammation and a dysentery-like illness.

Clinical features
- EPEC and ETEC—acute-onset watery diarrhoea without visible or microscopic blood and pus. Fever is usually absent. Usually short-lasting but EPEC may cause severe and protracted diarrhoea in hospital outbreaks
- EAggEC—acute-onset watery diarrhoea, commonly persisting beyond 14 days
- EIEC—fever, abdominal pains and watery stools becoming small volume, mucoid and bloody. Short-lasting
- After recovery excretion of organisms may be prolonged.

Diagnosis
E. coli are normally present in the human colon and these pathogenic strains do not normally differ from non-pathogenic strain in colony morphology. Colony samples can be further examined for suspected pathogenic strains by:
- EPEC—serotyping of known pathogenic strains (e.g. O119, O114, etc.), detection of adherence factor by DNA probe
- ETEC—demonstration of enterotoxin by immunoassay or of toxin gene by DNA probe
- EIEC—demonstration of enteroinvasive plasmid by DNA probe.

Treatment, prevention and prognosis
- Most cases require oral rehydration only
- In severe protracted diarrhoea co-trimoxazole may be effective
- In institutions infants should be nursed under strict enteric precautions
- Protracted cases of EPEC have a high fatality
- A single dose of ciprofloxacin and loperamide at the onset of symptoms is often effective for traveller's diarrhoea.

Cholera
Epidemiology and pathogenesis
- Caused by Gram-negative motile curved bacilli belonging to the species *Vibrio cholerae*
- Of the 139 known 'O' serogroups, only the serogroup O1 (classical and Eltor biotypes) and O139 are known to cause epidemic cholera

- Six previous pandemics have occurred during the past two centuries, all due to the 'classical' biotype originating in the Gangetic delta
- The seventh pandemic, currently in progress, is caused by biotype Eltor and began in 1961 in Indonesia, then spread across the world affecting large parts of Asia, Africa and South America
- Pockets of 'classical' cholera exist in the Gangetic delta
- In 1992, a previously unknown serogroup, O139 (*V. cholerae* Bengal), began causing epidemics of cholera in India and Bangladesh and has since spread to many countries of southeast Asia
- In nature, *V. cholerae* exists in aquatic environments, surviving in brackish estuaries and coastal waters in association with zooplankton. Environmental reservoirs of serogroup O1 exist in the Gulf coast of the USA and in Queensland, Australia
- Humans are the only natural hosts of *V. cholerae*
- Transmission is through consumption of contaminated food and water
- Children are more commonly affected
- Asymptomatic infections are not uncommon
- The IP is 2–3 days.

Pathogenesis

- Diarrhoea results from the action of cholera toxin (an enterotoxin) through stimulation of the adenylate cyclase/cyclic AMP pathway and possibly through other secretory factors
- Local intestinal immunity against the specific serogroup follows recovery.

Clinical features

- Sudden onset of vomiting and profuse watery diarrhoea with abdominal cramps is characteristic. Fever is absent
- Stools are white-yellow with mucus flecks ('rice water'), and the daily volume often reaches several litres
- Rapid dehydration leading to shock and death may occur within hours
- Mild and inapparent cases also occur
- Diarrhoea subsides and resolves within a few days

- Convalescent faecal excretion of organisms often lasts for several weeks.

Diagnosis

- Rapid presumptive diagnosis can be obtained by dark-field microscopy showing motile vibrios
- Stool culture is necessary for distinguishing cholera from other causes of watery diarrhoea
- DNA probes and PCR-based diagnostic kits have been developed.

Treatment

- The mainstay of treatment is early use of oral rehydration supplement (ORS) to correct fluid and electrolyte loss
- Severely dehydrated patients must receive intravenous (IV) fluids
- Ciprofloxacin or norfloxacin are effective in shortening the duration of the clinical illness, eliminating the vibrio from faeces (though resistance is increasing).

Prognosis

- Death rates may reach 50% when cholera strikes a previously unaffected and unprepared country
- Organized diarrhoea disease control programmes reduce fatality to 1%
- Most deaths are from shock or renal failure.

Prevention

- During an epidemic the following measures are essential:
  - hygienic disposal of human faeces
  - adequate supply of safe drinking water
  - good food hygiene
  - good personal hygiene
- Parenteral, killed whole-cell cholera vaccines give poor, short-lasting protection and are of little practical value (not available in the UK)
- Limited stocks of two oral vaccines that provide high-level protection for several months against *V. cholerae* O1 have recently become available in several countries and are suitable for use by travellers
- Vaccination certificate is no longer a travel requirement for any country.

Diarrhoea caused by other vibrios

Non-O1 V. cholerae

• Serogroups of *V. cholerae* other than O1 and O139 can cause sporadic acute diarrhoea in many countries

• These organisms are commonly found in environmental waters and transmission is through consumption of raw or undercooked seafood

• Severity of diarrhoea varies from mild to severe cholera-like.

Vibrio parahaemolyticus

• These organisms are widely distributed in coastal waters

• They are a common cause of both sporadic cases and outbreaks of diarrhoea in many countries, but particularly in Japan and other Far East/south-east Asian countries

• Raw or undercooked seafoods are the main vehicle of transmission

• The IP is 4 h to 4 days

• Acute watery diarrhoea with abdominal cramps is the most common manifestation, but bloody stools and high fever may also occur

• Treatment consists of maintenance of hydration and antibiotic (fluoroquinolone) in severe cases.

Clostridium difficile diarrhoea

Epidemiology

• *C. difficile* is the most important cause of antibiotic-associated diarrhoea worldwide

• These large, Gram-positive anaerobic bacilli with terminal spores are found in stools of about 3% of healthy adults and 10–20% of hospital patients

• Institutional outbreaks are common. Sporadic cases occur in the community

• Transmission is faecal–oral from direct contact between patients, or from the environment and from health-care workers. Instruments (e.g. sigmoidoscope) may be involved in transmission

• A predisposing factor is antibiotic use in the previous 4 weeks. Almost all antibiotics are reported to be associated but clindamycin has the highest incidence (per-use basis)

• The elderly are at greater risk.

Pathogenesis

• *C. difficile* produces two types of exotoxin, enterotoxin A and cytotoxin B, which can cause inflammation of colonic mucosa and fluid secretion

• Antibiotic therapy disturbs normal gut flora and allows *C. difficile* to flourish and produce toxins.

Clinical features

• Diarrhoea commonly begins 4–10 days (up to 4 weeks) after initiation of antibiotic therapy

• The stools are usually watery. White mucoid material is commonly present. Blood is rare

• Severity varies from mild to prominent diarrhoea

• Fever, abdominal pain and tenderness, and leucocytosis are common in severe cases.

Complications

• Hypoalbuminaemia and electrolyte disturbances

• Toxic megacolon or perforation.

Diagnosis

• Demonstration of *C. difficile* toxin in stool specimen

• Sigmoidoscopic changes vary from normal-looking mucosa, to various grades of inflammation: from mild erythema to multiple discrete, yellow-white plaques, which may become confluent (pseudomembranous colitis, PMC)

• Rectal biopsy shows glandular dilatation, epithelial loss and outpouring of mucus, fibrin and neutrophils on to the surface.

Treatment and prevention

• In hospital, infected patients should be nursed under strict enteric precautions

• Mild cases may only need discontinuation of the offending antibiotic. In others, oral vancomycin or metronidazole are effective

• Relapses may occur following either treatment regimens but respond to the same agent

• Multiple recurrences are difficult to treat

• Cholestyramine resin (to bind the toxin) may be tried; role of biotherapy is undecided

- Antimotility drugs should be avoided as they can precipitate toxic megacolon.

Prognosis
In elderly and debilitated patients with PMC the death rate may be 30%.

Yersinia enterocolitica **infection**
Epidemiology
- The Gram-negative organisms belong to the genus *Yersinia* and are distributed worldwide
- Most human infections are caused by a few serotypes, O:3, O:5, O:8 and O:9, which cause primarily zoonotic infections of domestic and wild animals and are often asymptomatic
- O:3 and O:9 serotypes are commoner in Europe, O:3 and O:8 in the USA
- *Y. enterocolytica* ranks after *Campylobacter*, *Salmonella* and *Shigella* in frequency of faecal isolation in Europe and America but clinical case reports are less common
- Scandinavian countries report a higher incidence of clinical cases than other developed countries
- Transmission is faecal–oral from contaminated food, milk or water, or contact with infected animals or patients. Water-borne and milk-borne outbreaks have occurred
- Highest incidence is among children
- The IP is 5 days (range 1–11 days).

Pathogenesis
After invading the small intestinal mucosa the organisms multiply in the reticuloendothelial cells of Peyer's patches, then migrate to the regional lymph nodes for further multiplication, causing:
- ileitis and less commonly colitis and mesenteric adenitis
- possible bloodstream invasion in patients with iron overload or immune deficiency.

Clinical features
- Acute febrile watery diarrhoea is normally the sole manifestation in young children
- In older children and adults, right lower quadrant abdominal pain may be present as well, mimicking appendicitis (due to underlying ileitis or mesenteric adenitis)

- Spontaneous recovery usually occurs within a few days.

Complications
- Septicaemia (in patients with iron overload or immune deficiency)
- Reactive arthritis (particularly in individuals with HLA-B27 antigen)
- Erythema nodosum.

Diagnosis and treatment
- Isolation of the organism from faeces or by serology
- Most cases are mild and self-limiting but severe cases should be treated with cotrimoxazole or ciprofloxacin.

Rotavirus gastroenteritis
Epidemiology
- Rotavirus (RNA virus) is the commonest cause of gastroenteritis throughout the world
- Most are caused by group A, and much less frequently by group B and group C, which occur in animals; other mammalian and avian rotaviruses do not appear to cause human infections
- Group A can be subdivided into a number of serotypes
- Highest incidence is in children between 6 and 24 months of age, though it is not infrequent in children up to the age of 4. Subclinical cases are common
- The virus can infect adults and neonates, though the disease tends to be asymptomatic or mild
- In temperate climates, almost all cases occur in winter. In warmer countries, cases occur throughout the year
- Nosocomial outbreaks in hospitals among newborns and infants are common
- Transmission is usually faecal–oral but may be faecorespiratory. The virus is stable in the environment and can be transmitted through water or contaminated surfaces
- The IP is 2–3 days.

Pathogenesis
- The virus multiplies within the proximal intestinal mucosa, damaging the microvilli and

the columnar epithelium which is replaced by immature cubiodal cells
• Fluid absorption is affected and osmotic diarrhoea results
• Immunity (local, humoral and cellular) develops after recovery, so repeat infections tend to be milder.

Clinical features
• Illness begins abruptly with fever, vomiting and watery diarrhoea of varying severity
• Symptoms usually cease after 4–6 days but may be protracted and severe in the immunocompromised
• Virus excretion ceases after a week or so but may be prolonged in the immunocompromised.

Diagnosis
• Usually by rapid antigen detection by enzyme immunoassay or immunochromatography
• PCR-based tests increase detection rates but are less rapid and not routinely available.

Treatment and prognosis
• Oral rehydration
• Although deaths are rare in the developed countries, 2–3% require hospitalization for rehydration
• 600 000 children die annually worldwide, most in the developing countries.

Prevention
• Hospitalized children should be nursed under enteric precautions
• No chemotherapeutic or chemoprophylactic agent is currently available
• Several promising vaccine candidates are under evaluation.

Enteric adenoviruses
Epidemiology
• Account for up to 10% of community-acquired diarrhoeas in young children; role in adults unclear
• Unlike the adenoviruses which cause respiratory infections, the enteric adenoviruses are difficult to propagate in tissue cultures but are identifiable in stools by electron micrography (EM)

• Types 40 and 41 are mostly involved
• They do not have a seasonal preponderance
• Transmission is faecal-oral
• The IP is 8–10 days.

Clinical features, diagnosis and treatment
• Asymptomatic infections are common
• Diarrhoea tends to be milder but somewhat more prolonged compared to rotavirus gastroenteritis
• Respiratory symptoms are rarely observed
• Diagnosis is by antigen detection or PCR
• Treatment involves attention to hydration.

Norwalk-like virus–associated gastroenteritis (winter vomiting disease)
Epidemiology
• Norwalk-like viruses (NLVs) account for most non-bacterial outbreaks worldwide
• They are small round structured viruses (SRSVs) and belong to the family of human Caliciviridae
• Seroepidemiology studies show evidence of past infection in 60% of adults
• In the developed countries older children and adults are most commonly affected, the viruses accounting for 5–17% of cases of community-acquired diarrhoeas and possibly accounting for a significant number of infantile diarrhoeas
• In the developing countries infections occur earlier
• Both sporadic infections and outbreaks are common, though clinical cases are usually diagnosed only in outbreaks
• Outbreaks typically occur in holiday and army camps, cruise ships, hotels and institutions, within families and especially in crowded and unsanitary conditions
• Often the outbreak grumbles along through introduction of new susceptibles to the infected area
• Humans are the only reservoir of infection
• Transmission is usually faecal–oral by food (commonly salads and vegetables), or water and ice contaminated by the faeces of an infected

person; consumption of shellfish harvested from or swimming in sewage-polluted water; or directly from person to person
- Aerosolized vomitus and fomite transmission might facilitate spread during outbreaks
- NLVs require a low infectious dose (<100 virions) and are therefore highly infectious
- They can resist chlorination and extremes of temperature, hence are stable in the environment
- The IP is 12–48h.

Pathogenesis
- The virus causes blunting of jejunal villi
- The mechanism of diarrhoea production is not understood
- Immunity to future infections is incomplete and short-lasting.

Clinical features
- Illness begins abruptly with nausea, abdominal cramps and vomiting, often projectile
- Diarrhoea may be absent, mild or severe
- Headache and myalgia are common; mild fever is present in 50%
- Symptoms usually subside within 48h
- New information has shown that virus shedding continues for up to 2 weeks after recovery and not for 48–72h (the basis for the current exclusion advice during outbreak control).

Diagnosis and treatment
- Demonstration of NLVs in stool specimens is difficult and currently many outbreaks are diagnosed on clinical and epidemiological grounds, and absence of positive stool culture
- Specific diagnosis requires identification of viral RNA by PCR, or of viral particle by EM or ELISA antigen capture assay
- Development of simple and sensitive detection techniques are awaited
- Illness is short and requires only attention to hydration.

Prevention
Outbreak control measures:
- limiting contact between ill and well persons
- measures to prevent water-borne, food-

borne, fomite-borne and person-to-person spread within the outbreak area as transmission is often multimodal
- exclusion of patients from food-handling for 48h after recovery: this is often successful, although latest evidence suggests virus shedding may continue for 2 weeks, so affected persons should continue strict personal hygiene when allowed to return to work
- avoidance of new susceptibles in the outbreak area.

Giardiasis
Epidemiology
Diarrhoea due to *Giardia lamblia* (*intestinalis*), a flagellated protozoon, occurs worldwide.
- In the developed countries cases are usually seen in daycare centres and schools, among inmates of institutions with poor personal hygiene, among campers, in male homosexuals, in travellers to the developing countries and during water-borne community outbreaks
- In the UK, over 50% of cases are in 15–44-year age group
- In the developing countries it is very common generally. Prevalence rates can reach 20–30%
- The source of infection is usually humans, although animals such as beavers and other wild and domestic animals are possible sources
- Transmission is by faecal–oral transfer of cysts (not trophozoites) through person-to-person contact or consumption of infected food or water
- The cysts are hardy, may survive routine chlorination and faulty filtration and may live in cold water for months
- Infecting dose is low
- The IP is about 2 weeks.

Pathogenesis
- Each cyst releases two trophozoites in the upper gut which adhere to mucosa and multiply by binary fission
- Trophozoites change to cysts in colon, and are then excreted
- Mechanism of acute diarrhoea is unknown. Invasion or structural alterations are usually absent

• In chronic infection with malabsorption, partial villous atrophy is often found
• Immune mechanisms involved in recovery are poorly understood
• Infection in agammaglobinaemics is often severe and prolonged but not in cellular immunodeficient persons.

Clinical features
• Many infections are asymptomatic
• Illness may present abruptly with watery diarrhoea
• Flatulence, abdominal bloating and nausea are common and may predominate
• Symptoms often settle after a week
• Less severe diarrhoea and abdominal symptoms may persist continuously or inter-mittently for weeks or months
• In neglected cases malabsorption with steat-orrhoea and weight loss develop
• Untreated individuals may excrete cysts for long periods even if symptom free.

Diagnosis
• By identifying cysts or trophozoites in faeces by direct microscopy
• Multiple samples on separate days should be examined as cyst excretion is variable
• Trophozoites may be found in duodenal fluid sample (Enterotest or aspirate or biopsy)
• Also available: antigen detection by immunoassay and parasite demonstration by immunofluoresence.

Treatment
• Metronidazole (2 g daily for 3 days) or tinida-zole (2 g single dose) are effective in around 90%
• Relapses may occur but usually respond to same treatment. Multiple courses may be necessary. Investigate for reinfection
• Mepacrine (unlicensed in UK) is also effective, but side-effects are common.

Prevention
See p. 120.

Cryptosporidiosis
Epidemiology
• The coccidian protozoon *Cryptosporidium*

parvum is a major cause of human diarrhoea worldwide
• It accounts for up to 5% of diarrhoea cases in the developed world. Young children and their carers, handlers of farm animals, male homo-sexuals and travellers are more commonly affected
• The prevalence is significantly higher in the developing countries
• The organisms are widely prevalent in the animal kingdom, causing both disease and asymptomatic carriage
• Oocysts in faeces are ingested by another host, animal or human, infect the small intestinal mucosa and multiply asexually by schizogony, then sexually into oocysts, which are again excreted in faeces
• The oocysts are very hardy, survive in the en-vironment for long periods and resist routine chlorination
• Transmission is either faecal–oral (person to person, animal to person) or by means of contaminated food and water. Contaminated natural water is an important source of infection for travellers to developing countries. Outbreaks occur worldwide from faulty public water supplies (chlorination only or faulty filtration)
• Outbreaks in nurseries and schools occur and during farm visits. Attack rate is high in nurseries
• Two epidemiologically distinct genotypes exist: genotype 1 generally causes person-to-person infections (direct transmission or through contaminated water); and genotype 2—when the source is an animal or a human
• The IP is usually 7 days.

Pathogenesis
• The organisms multiply in the upper gut mucosa but the mechanism of the secretory diarrhoea is unknown
• In severely immunocompromised AIDS patients, chronic infection often results and the whole intestine may become involved, with extension into the biliary tract causing scleros-ing cholangitis or cholecystitis
• Cellular immunity is involved in recovery, though the exact mechanism is not clear.

Clinical features

- Asymptomatic infection may occur
- In others, watery diarrhoea and abdominal discomfort begin abruptly
- Fever and vomiting may occur
- In immunocompetent hosts, diarrhoea usually subsides within 14 days but may last for 30 days. Faecal excretion of oocysts continues for 2–4 weeks from onset
- In cellular immunodeficient hosts, especially in AIDS patients, chronic infection with profuse watery diarrhoea often continues, causing dehydration and malnutrition, and death rate is high (see p. 177).

Diagnosis

- Demonstration of acid-fast oocysts in the faecal smear, using modified acid-fast stains or immunofluorescent methods
- Numerous cryptococci are seen in EM of rectal biopsies in chronic cryptosporidiosis of AIDS (Fig. 9.2)
- Genotyping is useful in epidemiological investigations.

Treatment

- Attention to hydration
- No effective therapy is available.

Prevention

See p. 120.

Cyclosporiasis

Epidemiology

- *Cyclospora cayetanensis* is a recently iden-tified cause of human diarrhoea world-wide
- It is most common in tropical and subtropical countries with outbreaks in Nepal, India, south-east Asia and South America
- Outbreaks have occurred in the USA, Canada and Europe, usually linked to imported fresh vegetables from endemic countries and travellers to these areas
- Transmission is by means of food or water contaminated by faeces of humans and possibly birds or other animals
- Oocysts can adhere stubbornly to vegetables despite washing
- The IP is probably very short — 1–2 days.

Pathogenesis

- After ingestion, the sporulated oocysts invade the small intestinal mucosal cells and multiply asexually, then sexually to form oocysts, which are shed faecally
- Freshly passed oocysts are not infectious (hence lack of direct faecal–oral transmission)
- Sporulation into infectious form occurs outside in weeks or months
- Mechanism of diarrhoea is unknown.

Clinical features

- Watery diarrhoea and abdominal discomfort begin abruptly
- Diarrhoea is often prolonged or waxes and wanes for several weeks with flatulence, bloating, dyspepsia and weight loss
- These symptoms are commonly more protracted in AIDS patients.

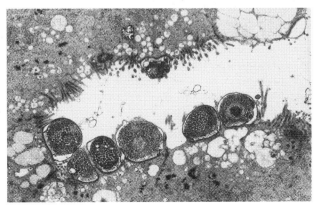

Fig. 9.2 Electron micrograph of rectal biopsy showing cryptococci attached to mucosa in an AIDS patient.

Diagnosis

- Is by demonstrating oocysts in faeces by microscopy
- These are acid-fast round bodies, larger in diameter than cryptosporidium
- They may also be found in jejunal aspirate or biopsy (EM).

Treatment and prevention

- Co-trimoxazole shortens the duration of illness
- AIDS patients may require maintenance treatment
- See p. 120 for prevention.

Amoebiasis

Epidemiology

- Several species of the protozoan genus *Entamoeba* infect humans, but only *E. histolytica* causes disease
- Humans are the only reservoir
- It occurs widely throughout tropical and subtropical Asia, Africa, the Middle East, and Central and South America, particularly in poor socioeconomic conditions
- In the developed countries, it is seen in recent immigrants from and travellers to the above, in mental institutions and in male homosexuals
- Transmission is by ingestion of cysts in faecally contaminated food or water
- Direct person-to-person transmission occurs among male homosexuals
- In the small intestine the cyst is digested, releasing four trophozoites, which migrate downwards to live on the colonic surface and multiply by binary fission
- Encystation occurs in the left side of the colon where the faecal contents are more solid
- Cysts can survive outside the human body for a long time, particularly in moist conditions
- Trophozoites do not transmit infection
- 90% of infected persons are asymptomatic cyst passers
- The IP is 2–6 weeks.

Pathogenesis

- Only trophozoites can invade tissues and a number of virulence factors have been identified

- Following attachment to intestinal mucosa trophozoites cause inflammation and ulceration that extends underneath viable mucosa causing the characteristic flask-shaped ulcers
- Invasive trophozoites typically ingest red cells
- Sometimes, a segment of ileocaecum may become diffusely thickened from chronic inflammation, presenting as a right lower abdominal mass (amoeboma)
- Trophozoites may invade local veins and migrate via the portal venous circulation to the liver (usually right lobe) and produce abscess
  - corticosteroids predispose to blood stream invasion
  - invasion results in hepatocyte necrosis which replaces liver parenchyma; neutrophilic infiltration is minimal
  - the necrotic material is like 'anchovy sauce' and is surrounded by a thin layer of congested viable liver tissue, the so-called capsule
  - trophozoites are found only in the capsule
- Immunity to future invasive disease develops following recovery but not to fresh colonization of intestines.

Clinical features

- Gradually worsening bloody mucoid diarrhoea over 1–3 weeks, cramp-like abdominal pains and variable pyrexia
- The symptoms usually persist for several weeks before subsiding, even if no treatment is given
- However, relapses are common, occurring irregularly over months or years
- Cyst passing may continue for years in the untreated, even if asymptomatic.

Complications

- Fulminant colitis in patients who are immunocompromised, pregnant or otherwise debilitated and malnourished
- Amoeboma
- 'Chronic amoebiasis': in endemic areas, cysts are not uncommon in individuals with intermittent, non-bloody but mucoid diarrhoea. A few have low-grade invasive amoebic infection, but most have asymptomatic infection with unrelated diarrhoea

- Amoebic liver abscess (commonly presents months after subsidence of diarrhoea):
  - often presents acutely with short-duration fever, and dull or pleuritic right upper quadrant pain
  - subacute presentation with weight loss and hepatomegaly of many weeks' duration may occur
  - fever may be the only symptom
  - right-sided pleural effusion is common
  - jaundice is rare
  - may rupture into pleural space and then into lung with patient coughing up anchovy-coloured sputum; or into peritoneum or pericardium.

Prognosis
- It is the third leading parasitic cause of death after malaria and schistosomiasis
- 40 000–100 000 deaths occur annually in the developing countries
- Less than 1% die from liver abscess if diagnosed and treated early.

Diagnosis
- Sigmoidoscopy may suggest the diagnosis by revealing shallow ulcers surrounded by normal-looking rectal mucosa. However, in more severe infections, there may be diffuse haemorrhagic colitis
- Confirmation is by demonstrating motile trophozoites, containing red cells, in fresh stool sample

- Trophozoites may be seen in the rectal biopsy specimens
- Positive serology is useful in the presence of a suggestive clinical picture but may be positive in asymptomatic infections. Usually becomes negative within a year
- See p. 141 for differential diagnosis of amoebic liver abscess.

Treatment and prevention
- Metronidazole (effective against trophozoites) is the drug of choice for active colonic disease and for liver abscess
- Aspiration of liver abscess is required only if rupture is likely or when distinction from pyogenic abscess is necessary
- A 10-day follow-up course of diloxanide furoate is necessary to destroy colonic cysts
- In non-endemic countries, asymptomatic cyst excretion should be treated with diloxanide furoate to prevent future disease
- See p. 120 for prevention.

Outbreaks of food-borne infections
- An outbreak occurs when two or more people consuming the same food develop identical illness, usually gastrointestinal, sometimes neurological
- Most commonly recognized pathogens in the developed countries are *Salmonella, C. perfringens, E. coli* O157 and NLVs
- Pathogens vary according to the food involved (see box below).

FOOD INVOLVED AND THE LIKELY PATHOGEN

Food	Pathogen
Red meat	Clostridium perfringens, Salmonella, Escherichia coli O157
Poultry and eggs	Salmonella, Campylobacter
Fish and shellfish	NLVs, scrombotoxic fish poisoning, shellfish poisoning, ciguatera poisoning
Salads and vegetables	NLVs, Salmonella, Cryptosporidium
Milk	Campylobacter, Salmonella
Potato salads, mayonnaise, pastries	Staphylococcus aureus
Fried rice	Bacillus cereus
Canned food	Clostridium botulinum

Investigation of outbreaks of food-poisoning

- Speedily review all suspected cases with regard to time, place and everybody at risk
- List all foods eaten in the previous 72 h
- Obtain and refrigerate samples of any such foods still available
- Determine on clinical grounds the likely pathogens and alert the investigating laboratory accordingly
- IP is often a useful guide: short periods of less than 6 h suggest ingestion of preformed toxin (Table 9.3), whereas IPs greater than 12 h, the presence of fever and symptoms continuing beyond 24 h suggest actual infection by living organisms
- Institute infection control procedures in institutions (e.g. hospitals, camps)
- Obtain samples of vomit and faeces and have them examined for bacterial and viral causes
- Follow carefully the food chain from supply to kitchen to table, to determine any unsafe practices
- Perform case-control or cohort study to analyse the association between the suspected foods and disease.

Typhoid and paratyphoid fevers (enteric fever)

Epidemiology

- The causative organisms *S. typhi* and *S. paratyphi* A, B and C, belong to the genus *Salmonella,* and are human pathogens
- The infections are most prevalent in south and south-east Asia, the Middle East, Central and South America and Africa. A low level of endemicity exists in south and eastern Europe (mostly of paratyphoid B)
- Multidrug resistance to chloramphenicol, ampicillin and co-trimoxazole is common in south-east Asia
- An epidemic of quinolone-resistant typhoid has occurred in Tajikistan
- In the developed countries, enteric fever is largely an imported infection (around 200 cases of typhoid and 150 cases of paratyphoid fevers are seen in the UK annually)

- Paratyphoid C is rare everywhere
- Transmission is through food or water contaminated by the faeces or urine of a patient or carrier. Direct case-to-case spread is uncommon
- The IP is 10–21 days.

Pathogenesis

- The organisms penetrate the intestinal mucosa and travel to the regional glands to multiply, then enter the bloodstream in large numbers, marking the onset of fever
- The Peyer's patches of the ileum are infected during this bacteraemia and also later through infected bile
- They become inflamed and later during the 2nd or 3rd week of illness may ulcerate, causing haemorrhage and perforation
- The liver and gall bladder are also involved
- After recovery, infection may persist indefinitely in the biliary and urinary tracts, particularly in the presence of pre-existing disease, leading to chronic faecal or urinary carriage
- After recovery, local intestinal, cellular and humoral immunities develop and second attacks are rare.

Clinical features

- Untreated typhoid fever is often a severe prolonged illness lasting for 4 weeks or more.
  - *First week:* mounting fever, headache, malaise, constipation, unproductive cough, relative bradycardia
  - *Second week:* continuous fever, apathy, diarrhoea, abdominal distension, 'rose spots' (in 30%), splenomegaly (in 75%)
  - *Third week:* continuous fever, delirium, drowsiness, gross abdominal distension, 'pea-soup' diarrhoea
  - *Fourth week:* gradual improvement in all symptoms
- After recovery, relapse may occur in up to 10% of cases (rare after fluoroquinolone therapy)
- Mild and inapparent cases occur
- Paratyphoid cases are similar to typhoid but generally milder.

Table 9.3 Microbial food-poisoning—toxin associated.

Type	Occurrence and source	Pathogenicity	Clinical features
Staphylococcus aureus food-poisoning	Wordwide Pastries, milk products	Multiplying organisms in unrefrigerated food produce heat-stable enterotoxin which is not destroyed during reheating	IP: 2–4 h (range 1–7) Nausea, vomiting, abdominal pain, diarrhoea rare, fever absent, settles rapidly
Clostridium perfringens food-poisoning	Commonest red meat-associated food-poisoning in the West Organisms inhabit human and animal intestines Beef and lamb (large joints), less commonly poultry	Heat-resistant spores survive cooking and germinate during reheating Produce heat-labile enterotoxin in intestine after ingestion	IP: 12–18 h (range 8–22) Diarrhoea, abdominal pain Vomiting rare, fever absent, settles <24 h
Bacillus cereus food-poisoning	Worldwide The organisms are ubiquitous in soil and contaminate rice and other cereals Fried rice most commonly involved	Heat-resistant spores survive boiling of rice and germinate if left in room temperature A heat-stable enterotoxin produced during storage survives quick frying and causes the emetic form of illness or A heat-labile enterotoxin is produced *after* ingestion and produces the diarrhoeal illness	Emetic form — IP: 1–5 h. Predominantly upper GI symptoms Diarrhoetic form — Predominantly diarrhoea Fever absent, settles <24 h

Continued on p. 138

Table 9.3 (continued)

Botulism (see p. 85)			
Scrombotoxic food-poisoning	Commonest form of toxin-associated fish poisoning in UK and USA, occurs worldwide. Scromboid (tuna, mackerel) and rarely non-scromboid (sardines, pilchards) fish	Caused by bacterial spoilage of fish allowing accumulation of protein breakdown products like histamine	IP: <2 h Flushing, headache, nausea, palpitations Settles within hours
Fish and shellfish food-poisoning from contaminated marine algal toxin	Occurs worldwide. Toxins produced during warmer months enter and concentrate inside shellfish during filter feeding, or in smaller fish and concentrate up the food chain in bigger fish feeding on smaller ones i.e. predatory tropical reef fish (ciguatera poisoning)	Toxins are heat stable, not destroyed by cooking or processing and produce distinctive clinical syndromes. Identified as diarrhoetic shellfish poisoning (DSP), paralytic (PSP), amnesic (ASP), neurotoxic (NSP) and ciguatera poisoning	Onset: few minutes to 30 h DSP: diarrhoea, abdominal pains, vomiting PSP: nausea, vomiting, numbness of lips and limbs, death may occur from paralysis NSP: vomiting, abdominal pains, numbness in mouth and limbs, seizures, respiratory symptoms ASP: diarrhoea, abdominal cramps, vomiting, headache. Memory loss and death may occur Ciguatera poisoning: usually a combination of gastrointestinal and neurological symptoms

Complications
- Intestinal haemorrhage and perforation (mostly in 3rd week)
- Myocarditis
- Neuropsychiatric: psychosis, encephalo-myelitis
- Cholecystitis, cholangitis, hepatitis, pneumonia, pancreatitis
- Abscesses in spleen, bone or ovary (usually after recovery)
- Chronic carrier state (positive stool/urine culture after 3 months) in 3% (less after fluoroquinolone).

Investigations and diagnosis
- White cell count is normal or leucopenic. Leucocytosis occurs when there is haemorrhage or pyogenic complication
- Definitive diagnosis requires isolation from blood or bone marrow
- Blood culture is positive in 80% in 1st week, progressively less common thereafter or when there is prior antibiotic use
- Bone marrow culture may remain positive despite antibiotic administration
- Stool and urine cultures are often positive from 2nd week onwards, diagnostic only if clinical picture is compatible
- Measurement of 'O' and 'H antibodies (Widal test) is unreliable and often difficult to interpret in the previously immunized or infected with related salmonella groups, so not used in the West
- A number of newer, more sensitive serodiagnostic tests (e.g. Vi antibody test) and Vi antigen detection by PCR are under evaluation.

Treatment
- Antibiotic therapy: ciprofloxacin orally or IV for 10–14 days in adults or third-generation cephalosporin (e.g. ceftriaxone) in children
- Chloramphenicol is a cheaper alternative in areas where the organisms are still sensitive
- Adjunctive IV dexamethasone reduces mortality in severely toxic, obtunded patients
- 75% of chronic carriers can be cured by 28-day courses of ciprofloxacin or norfloxacin

- Cholecystectomy should be performed only when symptoms of gall bladder disease warrant this
- Surgery is necessary for perforation, but haemorrhage is managed conservatively
- In non-endemic countries, epidemiological investigations are necessary to identify the source of infection if not acquired abroad
- After clinical recovery, three stool and urine cultures should be obtained on separate days and if any is positive, monthly cultures should be obtained until there are three consecutive negatives or 12 months have elapsed
- Food handlers should stay off work until then.

Prognosis
- Untreated typhoid has a mortality rate approaching 20%
- Mortality is negligible with prompt treatment
- The high death rates still prevalent in many endemic countries are due to delayed or inappropriate treatments.

Prevention
- Individuals travelling to or living in highly endemic areas should receive typhoid vaccines. Three types are available, all giving about 70% protection for 3 years.
  - Killed whole-cell vaccine: two injections are necessary for primary course. Local and systemic side-effects are common. It is cheap
  - Vi capsular polysaccharide: single injection, minimal local and systemic reactions. Immune response suboptimal in children < 18 months. A conjugate Vi polysaccharide vaccine has given around 90% protection in children > 2 years and may be more suitable in infants
  - Ty 21a live attenuated oral vaccine: 3 capsules over a 5-day period. Virtually free of side-effects but costly. Not suitable for children < 5 years of age
- In typhoid-endemic countries the most important measures are provision of safe drinking water, safe disposal of excreta and public education on hygiene.

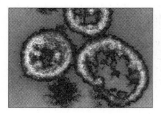

Infections of the Liver and Biliary Tract

Clinical syndromes

Liver abscess
Major presenting features: *fever, abdominal pain.*

Causes
Predisposing
• Biliary disease, gastrointestinal tract conditions (e.g. appendicitis, diverticulitis), systemic infection, contiguous infection
• Residence in or travel to an area endemic for amoebiasis.

Microbiological
• STREPTOCOCCUS MILLERI, ANAEROBES, *Escherichia coli*, other 'coliforms', *Enterococcus faecalis*, microaerophilic streptococci (often polymicrobial)
• STAPHYLOCOCCUS AUREUS, *Streptococcus pyogenes*, *Brucella suis*, actinomyces, *Salmonella typhi*, other salmonella
• ENTAMOEBA HISTOLYTICA
• *Mycobacterium tuberculosis*, hydatid disease, *Candida.*

Distinguishing features
See Table 10.1.

Complications
Extension: empyema, subphrenic abscess, hepatobronchial fistula, skin, intrahepatic obstruction of major bile duct.

Rupture: pericardium, peritoneum, bile duct, gastrointestinal tract.
Metastatic: lung abscess, brain abscess.
Others: pleural effusion, septicaemia (pyogenic abscess), secondary bacterial infection (amoebic liver abscess).

Investigations
• Full blood count (FBCt) (anaemia common), white cell count (WCCt) (raised) and erythrocyte sedimentation rate (ESR) (raised)
• Biochemical profile (liver function tests (LFTs), albumin and urea/creatinine). A raised alkaline phosphatase is often found
• Chest X-ray (CXR) (Fig. 10.1): raised hemidiaphragm anteriorly; pleural effusion/empyema; basal collapse; gas/fluid level in abscess; pericardial effusion (follows on from a left lobe abscess)
• Abdominal X-ray (calcification in the gall bladder or a hydatid cyst; gas/fluid level in abscess)
• Imaging: ultrasound scan; isotope scan (technetium/gallium); computed tomography (CT) scan (Fig. 10.2)
• Blood culture (positive in pyogenic abscess)
• Abscess aspirate: microscopy (for trophozoites/cells); Gram stain; culture; cytology (malignancy can mimic an abscess)
• Serology for *E. histolytica*
• Stool microscopy (for amoebic trophozoites/cysts).

	Pyogenic liver abscess	Amoebic liver abscess
Incubation period	1–16 weeks	1–12 weeks
Fever	High, hectic	Low grade
Toxicity	May be marked	Minimal or absent
Liver		
Tenderness	Usual	Invariable, may be intercostal
Swelling	Uncommon	Common
Jaundice	25%	5%
Clubbing	If chronic	Never
Preceding events	Intra-abdominal infection/surgery	Dysentery in 20%
Stool microscopy	Normal	E. histolytica cysts/trophozoites in 15%
Blood culture	Positive in 34%	Negative
Abscess aspirate		
Gram	Positive	Negative
Culture	Positive	Negative
Trophozoites	Negative	Occasionally positive
Amoebic serology	Negative	Positive
Abscess number	Multiple in 35%	Rarely multiple

Table 10.1 Distinguishing features of liver abscess.

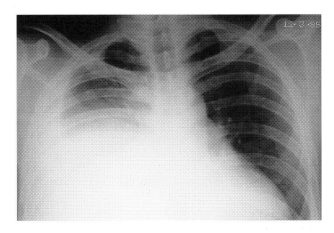

Fig. 10.1 Raised right hemidiaphragm from a large amoebic liver abscess just before it spontaneously ruptured into the lungs.

Differential diagnoses
Hepatoma, metastases, cholangiocarcinoma, hepatic cyst.

Treatment
Pyogenic liver abscess
• Aspiration/drainage percutaneously under ultrasound scan (USS) guidance
• Antibiotics: metronidazole (for anaerobes), ampicillin (for streptococci, enterococci) and cefotaxime (for 'coliforms', S. aureus); or co-amoxiclav (antibiotics may need modification after culture results available)
• Formal surgery.

Amoebic liver abscess
• Metronidazole
• Aspiration/drainage percutaneously under

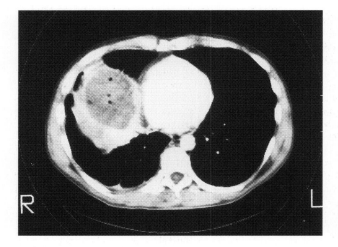

Fig. 10.2 A CT scan shows the location of the abscess (same patient as in Fig. 10.1).

USS guidance (if large abscess, left lobe, about to point, negative amoebic serology, doubt about amoebic aetiology)
• Luminal antiamoebicide — diloxanide furoate.

Prevention
• Prophylactic antibiotics for gastrointestinal surgery
• General travel advice concerning eating cooked food and boiling water.

Hepatitis
Major presenting features: *malaise, nausea/vomiting, jaundice.*

Causes
Acute hepatitis (transaminases 10–100× normal)
• HEPATITIS A VIRUS (HAV), HEPATITIS B VIRUS (HBV), hepatitis C virus (HCV), hepatitis delta virus (HDV), hepatitis E virus (HEV)
• Cytomegalovirus (CMV), Epstein–Barr virus (EBV)
• Yellow fever virus.

Acute hepatitis (transaminases 2–10× normal)
• Leptospirosis, *Coxiella burnetii, Mycoplasma pneumoniae*, brucellosis, *Legionella pneumophila, Chlamydia psittaci*
• Pneumococcal pneumonia, Gram-negative bacteraemia

• Arboviruses
• *Treponema pallidum*, toxoplasmosis, *Candida.*

Chronic hepatitis (transaminases 1–10× normal)
• HBV, HCV, HDV
• *Coxiella burnetii, M. tuberculosis.*

Distinguishing features
See Table 10.2.

Complications
Acute viral hepatitis
All types: fulminant hepatic failure, cholestatic hepatitis, relapsing hepatitis, aplastic anaemia, chronic fatigue syndrome.
HBV, HCV, HDV: chronic active hepatitis, cirrhosis.

Chronic viral hepatitis
HBV, HCV, HDV: cirrhosis.
HBV, HCV: hepatoma, vasculitis (polyarteritis nodosa (PAN), glomerulonephritis).

Leptospirosis
Renal failure, myocarditis, adult respiratory distress syndrome (ARDS) and disseminated intravascular coagulopathy (DIC) with fulminant progression.

Investigations
• FBC↑, differential WCC↑

	Acute viral hepatitis	Weil's disease (leptospirosis)
Incubation period	Short (2–8 weeks): HAV, HDV, HEV Long (1–6 months): HBV, HCV	10 days (6–15 days)
Onset	Gradual	Sudden
Risk history	Contact, travel, shellfish (HAV/HEV) Blood/sexual contact, IVDU, institutions (others)	Contact with animals/contaminated water
Fever	Normal/low	High
Headache	Occasional	Constant
Chest symptoms	Rare	Not uncommon
Myalgia	Mild	Severe
Toxaemia	Absent	Marked
Conjunctivae	Normal	Suffused
Bleeding	Rare	Not uncommon
Liver failure	May occur (acute/chronic)	Never
Proteinuria	Absent	Present
White cell count	Normal	Raised
Peak transaminase	100 × normal	2–5 × normal

Table 10.2 Distinguishing features of viral
hepatitis and leptospirosis.

• Clotting profile (prolonged clotting in liver
failure)
• Biochemical profile (LFTs, albumin and
urea/creatinine)
• Glucose (lowered in fulminant hepatitis)
• CXR (to exclude pneumonia, check for
ARDS).

Acute viral hepatitis
• Serology
• Initial screen:
  • HBV surface antigen (HBVsAg)
  • HAV immunoglobulin M (IgM) antibody
• If initial screen negative:
  • HCV antibody (consider HCV polymerase
  chain reaction (PCR))
  • heterophile antibody test (Monospot)
  • serology for EBV, CMV, *C. burnetii*, *M. pneu-
  moniae*, *Legionella pneumophila*, *Leptospira*,
  Chlamydia, *Toxoplasma* and syphilis
• If HBV surface antigen positive:
  • HBV IgM antibody (this confirms recent
  infection)

• HBV e antigen antibody (this determines
infectivity)
• HDV antibody (if history suggests possible
cause).

Chronic viral hepatitis
• Serology:
  • HBV surface antigen (HBVsAg)
  • HCV antibody
  • HDV antibody (if HBVsAg negative)
• To determine activity of disease:
  • nucleic acid detection (HBV DNA, HCV
  RNA, HDV RNA)
  • liver biopsy
  • genotype of HCV if HCV RNA positive
• α-fetal protein (screen for hepatoma)
• Ferritin or iron studies (screen for
haemachromatosis)
• Autoimmune screen (to exclude lupoid
hepatitis)
• Monitoring should include regular α-fetal
protein, USS and endoscopy if portal
hypertension.

Differential diagnoses

Drugs, alcohol, autoimmune (lupoid)
hepatitis, toxins, anoxic liver damage, Wilson's

disease, haemochromatosis, α_1-antitrypsin deficiency, veno-occlusive disease, granulomatous
hepatitis.

Treatment
Acute viral hepatitis
- Bed rest
- Full intensive care for fulminant cases
- Liver transplantation.

Chronic viral hepatitis
- HBV: α-interferon, lamivudine or adefovir
- HCV: pegylated α-interferon with ribavirin
- HDV: α-interferon
- Liver transplantation.

Prevention
Pre-exposure
- Passive immunization: gammaglobulin for
HAV (may be effective for HEV)
- Active immunization:
  - HAV vaccine: booster at 6–12 months
  gives 10-year protection
  - HBV vaccine: 0, 1, 6 months (serum
  antibody check at 6 weeks after last
  immunization).

Postexposure
- Passive immunization:
  - gammaglobulin for HAV
  - hyperimmune HBV globulin for HBV

Table 10.3 Distinguishing features of biliary tract
infection.

- Active immunization:
  - HAV and HBV (accelerated course at 0, 2, 4
  and 8 weeks).

General measures
See preventive measures listed under specific
diseases, later in this chapter.

Cross-reference
Leptospirosis (see p. 237).

Cholecystitis and cholangitis
Major presenting features: *fever, right upper
quadrant pain, jaundice.*

Causes
Cholecystitis
Predisposing
- Gallstones.

Cholangitis
Predisposing
- Gallstones, recent biliary surgery, biliary
stricture, pancreatitis, *Clonorchis* infection,
prior endoscopic retrograde cholangiopancreatography (ERCP), cholangiocarcinoma.

Microbiological
- ESCHERICHIA COLI, OTHER 'COLIFORMS', ANAEROBES, E.
FAECALIS, *S. typhi*, other salmonellae.
- Kawasaki disease.

Distinguishing features
See Table 10.3.

	Cholecystitis	Cholangitis
Preceding history	Gallstones	Gallstones, ERCP, surgery
Fever	Normal/low	High, hectic
Rigors	Absent	Present
Toxicity	Absent or mild	Marked
Jaundice	Rare	Present
Abdominal tenderness	Subcostal, on inspiration	Upper abdominal
Palpable gall bladder	40%	Uncommon
White cell count	Mild/moderate rise	Marked rise
Bilirubin	Normal/mild rise	Marked rise
Alkaline phosphatase	Normal/mild rise	Moderate rise
Positive blood culture	Rare	50%
Ultrasound of abdomen	Patent, undilated common bile duct	Dilated common bile duct

Complications

Cholecystitis: perforated gall bladder, biliary peritonitis, emphysematous cholecystitis, empyema of the gall bladder, cholangitis.

Cholangitis: septicaemia, septic shock, perforated gall bladder, liver abscess, pancreatitis, empyema of the gall bladder.

Investigations
- FBCt, differential WCCt, ESR
- Biochemical profile (LFTs, albumin and urea/creatinine): mildly obstructive pattern usually seen
- Cardiac enzymes and electrocardiogram (ECG) to exclude myocardial infarction (MI)
- Clotting profile (may be prolonged prothrombin time in cholangitis)
- CXR (to exclude pneumonia)
- Plain abdominal X-ray (for calcification, gas in gall bladder/common bile duct)
- USS of upper abdomen
- Blood culture.

Differential diagnoses
Viral hepatitis, Weil's disease, liver abscess, pancreatitis, metastatic liver disease, pyelonephritis, peptic ulcer, MI, lower lobe pneumonia.

Treatment
- Intravenous (IV) cefuroxime (for 'coliforms'), metronidazole (anaerobes) with or without ampicillin (*E. faecalis*); or IV co-amoxiclav
- Surgery: immediate cholecystectomy (only if complications arise); elective cholecystectomy >3 months after acute cholecystitis
- ERCP and sphincterotomy if CBD gallstone obstruction.

Prevention
- Prophylactic antibiotics for biliary surgery and ERCP
- Therapy of gallstones.

Specific infections

Hepatitis A
Epidemiology
HAV is a global disease of humans, particularly prevalent in developing nations.

- Transmission is faecal–oral through contaminated food or water
- Infectivity is greatest in the week before the prodrome and tails off after symptoms develop, becoming very low by the time jaundice develops
- Anicteric infections are commoner in children (10:1) than adults (1:1)
- There are approximately 10 000 cases/year in the UK with an incidence of $15/10^5$ population; 5% of cases are acquired abroad
- Transmission rates are higher where there is poor sanitation and overcrowding, amongst preschool groups and male homosexuals, and within institutions
- In developing countries most HAV acquisition occurs in childhood. In developed nations 20% of young adults have serological evidence of past infection; incidence and severity rise with age
- Large outbreaks have been described resulting from contaminated water, milk or food. Shellfish may become infected from sewage-contaminated seawater and become a vehicle for transmission
- Lifelong immunity follows an attack
- The incubation period (IP) is 28 days (range 14–42 days).

Pathology and pathogenesis
HAV is an RNA enterovirus. Following ingestion, the virus reaches the liver via entry through the oropharynx or upper intestine. Replication is restricted to the liver. The virus is cytopathic but host-mediated immune responses also contribute to the acute hepatocellular damage. The findings on liver biopsy are non-specific (focal necrosis, portal inflammation, ballooning, acidophilic bodies) and do not accurately distinguish HAV from other types of acute viral hepatitis. Severe hepatitis may be associated with massive necrosis.

Clinical features
Distinction of HAV from other types of viral hepatitis on clinical grounds is very difficult:
- the onset is usually gradual with low-grade fever, myalgia, upper abdominal discomfort,

anorexia, nausea and vomiting. Smoking becomes distasteful
• after 3–6 days the urine becomes dark, the faeces pale and jaundice appears. This lasts for 1–2 weeks; other symptoms improve with its appearance
• arthralgia and rash occur in up to 5% of cases
• hepatomegaly is found in nearly all patients and splenomegaly in 20%.

Complications
Hepatic
• Fulminant hepatitis
• Cholestatic hepatitis
• Relapsing hepatitis.

Extrahepatic
• Aplastic anaemia, haemolytic anaemia, thrombocytopenia
• Posthepatitis syndrome (chronic fatigue syndrome).
Fulminant hepatitis occurs in 0.1% of cases with a fatality rate of 20%; this increases with age. Cholestatic jaundice is a common problem but always resolves and remains benign. Relapsing hepatitis 1–4 months after recovery associated with a return of hepatitis occurs rarely.

Diagnosis
A detailed clinical history and examination will suggest the diagnosis. Suspicion of hepatitis will come from:
• history of exposure (e.g. other cases in the family or school)
• very raised transaminases (e.g. alanine aminotransferase (ALT) or aspartate aminotransferase (AST) > 1000 IU/L)
• prolonged prothrombin time.
Confirmation of HAV is through detecting HAV IgM antibody (positive for 12 weeks). The immunoglobulin G (IgG) antibody remains positive for life.

Treatment
• Bed rest is the mainstay of treatment
• In fulminant cases, intensive care is necessary
• Liver transplantation can be life saving.

Prevention
• Good sanitation and water supply, together with personal hygiene, and avoidance of food or water likely to be contaminated
• Isolation of patients is unnecessary
• Children who are close contacts should be kept off school if they develop a febrile illness
• Shellfish should be cultivated in sewage-free waters
• Passive immunization with normal immunoglobulin is effective and provides 3 months' protection. However, with the advent of a highly effective vaccine, it is rarely used. It is indicated for short-term travellers going to endemic areas, pregnant women and immunocompromised persons who are at risk of severe disease after close contact, and in health-care workers after occupational exposure
• Active immunization with a killed vaccine produces excellent immunity. It is indicated for travellers going to endemic areas, to abort outbreaks, and to protect health-care workers post exposure, or pre-exposure where there is considered to be a significant occupational risk. Two injections are given 6–12 months apart.

Prognosis
HAV is usually a mild illness. The fatality rate is 0.03% of patients in those aged under 55 and 1.5% in those aged over 64 years.

Hepatitis B
Epidemiology
Worldwide, HBV is a major cause of disease through chronic liver disease and hepatoma.
• There are 10 000 new HBV infections/year acquired in the UK: the lifetime risk of hepatitis B is 5%
• Anicteric infections are common (4 : 1)
• 5–10% fail to resolve the infection and become carriers. This is more likely in persons with defective immunity
• The estimated carriage rate in the UK is 0.1%. In certain areas of the world, the carriage rate may exceed 25% (Pacific islands, Thailand, Senegal), and in others approximate 5–10% (large areas of the Indian subcontinent, southeast Asia, Africa and eastern Europe). The esti-

mate is that nearly 200 million persons world-wide are carriers
• Transmission is by blood or body fluid through injection or exposure to mucous membranes. Infection can therefore be acquired from blood products, contaminated needles or other medical equipment, and lifestyle events such as tattooing. Infection may also be contracted sexually. In a significant minority, there is no identifiable source
• Virus can be identified in most body fluids; saliva, seminal fluid, breast milk and serous cavity fluids are the most important (e.g. ascites)
• Newborn infants of infected mothers can acquire infection at birth. For HBVeAg-positive mothers, the risk of transmission is 90% and if HBVeAg negative, 15%. This is the major mode of transmission in the Far East with childhood horizontal transmission also being important in Africa. These are the major reasons for high population carriage rates. A small proportion of children get infected in utero
• Infection and carriage rates are higher in closed groups where blood or other body fluids are injected, ingested or exposed to mucous membranes. Hence, institutionalized children with mental handicaps, haemodialysis patients, intravenous drug abusers (IVDUs) and men who have sex with men have higher carriage rates (5–20%). Outbreaks may occur in these groups and via infected surgeons and dentists
• Several genotypes exist that appear to have no influence on the disease process. Molecular variants are also common. These are strains that do not express certain proteins as a result of key genetic mutations. Precore mutants do not express the 'e' antigen but are recognized by high serum titres of HBV DNA. Disease progression occurs rapidly and strains are less responsive to α-interferon
• Dual infection with the delta agent (HDV) may occur and is a particular problem amongst injection drug users
• The IP is 2–6 months.

Pathology and pathogenesis
HBV is a DNA virus possessing a surface coat (surface antigen) and an inner core (core antigen). In acute hepatitis, liver biopsy reveals varying degrees of hepatocellular damage and inflammatory infiltrate. HBV antigens are expressed on the surface of hepatocytes and there is T-cell-mediated cellular reactivity against these: this is presumed to be the major cause of hepatocyte damage. HBV antigens have also been identified in non-hepatic sites and may represent a reservoir of infection that can reseed the liver after transplantation. Patients with hypogammaglobulinaemia can develop acute hepatitis, indicating that antibodies do not play a major role in liver damage. In chronic hepatitis, varying degrees of histological activity may be seen. Several systems for scoring severity exist which take in the key elements of inflammation, necrosis, and fibrosis (e.g. Metavir, Ishak and Knodell). This allows more objective treatment decisions to be made. Chronic hepatitis may range from very mild with minimal necroinflammation (portal zone lymphocytic inflammation but with no evidence of bridging necrosis or disturbed architecture) and no evidence of fibrosis (chronic persistent hepatitis, CPH), to very active disease with marked necroinflammation (chronic active hepatitis, CAH), to fully developed cirrhosis. Mild disease is generally non-progressive whereas active necroinflammation may develop into cirrhosis or hepatoma. Chronic hepatitis is associated with chronic hepatitis B carriage and viral integration into the chromosome.

HBV surface antigen: this is found on the surface of the virus and on the accompanying unattached spherical particles and tubular forms. Its presence indicates acute infection or chronic carriage (defined as >6 months). Antibodies to surface antigen will occur after natural infection or can be elicited by immunization.

HBV core antigen: this is found within the core of the virus but is not detectable in the blood. IgG core antibody remains positive after infection indefinitely and is therefore a marker of past naturally acquired infection. IgM core antibody is useful in distinguishing acute HBV infection from another form of hepatitis in an

HBV carrier (e.g. delta virus) and in the rare patient who clears their surface antigen quickly. IgM core antibody remains positive for 12 weeks.

HBV e antigen: this is part of the core antigen. It is found in acute infection and in some chronic carriers. Its presence is a marker of underlying viral activity and infectivity. Antibody to this antigen indicates a lower level of infectivity in chronic carriers.

HBV DNA: this parallels viral replication. It is found in acute hepatitis and chronic carriers with active disease.

Clinical features

The symptoms may be indistinguishable from HAV.

• The onset is usually insidious, with low-grade fever, anorexia, upper abdominal discomfort, nausea and vomiting, and distaste for cigarettes
• After 2–6 days, the urine darkens, the stools become paler and jaundice develops
• A syndrome of fever, arthralgia or arthritis, and urticarial or maculopapular rash occurs in 10% of patients before the onset of jaundice. In children this may be pronounced and is labelled papular acrodermatitis (Gianotti's syndrome)
• Smooth tender hepatomegaly is usual and splenomegaly occurs in 15% of cases
• Patients with chronic infection and active liver inflammation tend to suffer from mild constitutional symptoms such as fatigue. Where there is established cirrhosis, symptoms and stigmata of chronic liver disease may be observed.

Complications

Hepatic
• Fulminant hepatitis
• CAH, CPH, cirrhosis
• Cholestatic and relapsing hepatitis
• Hepatoma.

Extrahepatic
• Aplastic anaemia, haemolytic anaemia, thrombocytopenia
• Guillain–Barré syndrome (GBS), encephalomyelitis (rare)

• Posthepatitis syndrome (chronic fatigue syndrome)
• Glomerulonephritis, vasculitis.

In 90% of cases, the illness is benign and complete recovery ensues after 2–4 weeks. Fulminant hepatitis is more common with hepatitis B (1.0%) than with hepatitis A (0.1%) but remains rare; it is associated with infection with pre-S mutations in the genome of the HBV surface antigen and with acute coinfection and superinfection with delta virus (HDV). Chronic carriage occurs in 10% and is associated with mild (70%) or active (30%) chronic hepatitis on liver biopsy. Progression to cirrhosis or hepatoma occurs in 25–30% of chronic carriers; this is more likely to occur in patients with high levels of replication (e antigen carriers or high HBV DNA levels). HBV has been incriminated in some patients with membranous glomerulonephritis and vasculitis.

Diagnosis (Fig. 10.3)

A detailed clinical history and examination will suggest hepatitis.

• Suspicion of *acute HBV* will come from the demonstration of:
  • very raised transaminases (e.g. ALT or AST > 1000 IU/L)
  • prolonged prothrombin time
  • consistent exposure history
• Confirmation is by demonstrating HBV surface antigen
• Confirmation of *chronic HBV carriage* is by demonstrating:
  • HBV surface antigen > 6 months after acute infection
• Confirmation of *chronic HBV hepatitis* is by demonstrating:
  • raised transaminases > 6 months after acute infection
  • liver biopsy evidence of chronic hepatitis
  • exclusion of other causes of chronic liver disease.

The level of infectivity and predicting underlying pathological activity can be gauged by determining HBV e antigen and antibody, and HBV DNA.

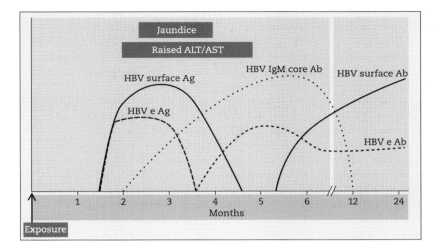

Fig. 10.3 Serological markers in uncomplicated acute hepatitis B infection. Ab, antibody; Ag, antigen.

Treatment

Acute hepatitis B
- Bed rest is the mainstay of treatment
- In fulminant cases intensive care is necessary
- Liver transplantation is complicated by the probability of reinfection of the graft from extrahepatic sites.

Chronic hepatitis B
Treatment can be either antiviral or through augmenting the immune system. Lamivudine and adefovir are licensed antiviral nucleoside drugs that have potent activity against HBV. Alpha-interferon enhances T-cell activity against infected hepatocytes. All agents have a success rate of approximately 30% in patients with good prognostic factors for response. Response is measured by conversion from HBVeAg positive to HBVeAb positive and reduction of HBV DNA to <200 copies/mL. Only about 10% lose HBVsAg. Response to interferon (20–30 megaunits per week for 4–6 months) is less likely in patients with normal transaminases, with high levels of HBV DNA, with minimal histological change, infected with precore mutant strain, who are of Asian origin, with evidence of immunocompromise, with a long period of carriage, and who are males. Interferon is associated with side-effects of influenza-like syndrome, marrow suppression, depression and alopecia: it should not be given to patients with established cirrhosis as worsening liver function may occur. Lamivudine is remarkably free of side-effects. It has the advantage of being equally active in precore mutant and wild virus strains and in Asians as well as non-Asians. It has the disadvantage that resistance develops in those who do not seroconvert at 1 year. Adefovir has similar antiviral potency but with a different site of action; resistance appears rare.

Liver transplantation for end-stage decompensated liver disease may be successful in selected patients, although liver reinfection from extrahepatic sites nearly always occurs. Suppression of viral replication at this time is important in protecting from post-transplant hepatitis.

Prevention
Two forms of protection are available: passive immunization with hyperimmunoglobulin to hepatitis B, and active immunization with vaccine.
- Vaccine is indicated for newborn children whose mother is HBV surface antigen positive

and for health-care workers post exposure if previously unimmunized. The standard vaccination schedule is at 0, 1 and 6 months: a booster is given to those who have not developed HBV surface antibody (HBVsAb) at 6–8 weeks after completion of the course. An accelerated course can be given in the postexposure situation (0, 2, 4 and 8 weeks)

• Hyperimmunoglobulin is indicated for newborn children of mothers carrying hepatitis B surface antigen who are also HBV e antigen positive or HBV e antibody (HBVeAb) negative. In the same instances it is also indicated for unimmunized health-care workers post exposure

• Routine immunization of at-risk groups is also important. These include all health-care workers, residents and staff of institutions for the mentally handicapped, and consorts and family members of a HBV e antigen positive carrier

• Elimination of high-risk persons as blood donors

• Screening of blood donors for HBV surface antigen

• Low-dose interferon has been shown to reduce the incidence of hepatoma in patients with cirrhosis.

Prognosis

Overall mortality of acute HBV is 1–3%, but 25–30% of patients with chronic carriage will have chronic hepatitis with necroinflammation, of whom 25% will develop cirrhosis and/or hepatoma. Median life expectancy after the onset of decompensated cirrhosis is less than 5 years and 1–3% develop hepatoma annually

Hepatitis C
Epidemiology

HCV is a single-stranded RNA virus which cannot be cultivated. Through recombinant DNA technology, a diagnostic test has been devised which has identified hepatitis C as the major cause of what used to be termed 'non-A non-B' (NANB) hepatitis.

• Six major genotypes exist: 1a, 1b, 2a, 2b and 3a are common in Europe and 4 in the Middle East and Africa (especially Egypt). Genotypes 2

and 3 respond significantly better to interferon therapy. Genotypes 5 and 6 are rare

• Studies in the 1970s and 1980s showed that the incidence of transfusion-associated NANB hepatitis was 10%; subsequently, HCV was identified in 90%

• The prevalence of HCV infection in the UK is 1–2/1000. It is estimated that there are 100 000 cases in the UK and 4 million in the USA. With the advent of effective treatment and recognition of the morbidity and mortality associated with hepatitis C, more patients are being screened. Consequently, it is a disease with a falling incidence but rising prevalence

• The major mode of transmission is through contaminated blood, most commonly via blood products (20%) or injection drug use (50%); 60% of users are HCV antibody positive

• Sexual, vertical and occupational transmission does occur but much less frequently than with HBV; this reflects the lower concentration of HCV in the blood

• 70% of anti-HCV positive persons have evidence of chronic hepatitis on biopsy but few have either signs or symptoms of liver disease

• 10–20% of patients with chronic hepatitis are likely to develop cirrhosis within 5–30 years. Once this is established, the incidence of hepatoma is 1% per annum

• Only 25% are icteric; many infections are asymptomatic

• The IP is 8 weeks (range 4–26 weeks).

Pathology and pathogenesis

The findings on liver biopsy in acute HCV are non-specific (lobular disarray, hepatocyte degeneration, lymphocytic infiltration). More severe hepatitis may be associated with massive necrosis. As in the case of HBV, two broad classes of chronic hepatitis have been defined—CPH and CAH—on the basis of the degree of necroinflammatory and fibrotic responses (see p. 147).

Clinical features

Clinically, acute HCV is indistinguishable from other causes of acute viral hepatitis. However, the majority of patients first present with

chronic infection (85%): fulminant infection is excessively rare.
- The onset of acute HCV is usually insidious, with low-grade fever, anorexia, upper abdominal discomfort, nausea and vomiting
- After 2–6 days, the urine darkens, the stools become paler and jaundice develops
- Prodromal arthralgia or arthritis may occasionally occur
- Smooth tender hepatomegaly is usual, and splenomegaly can occur.

In chronic disease, the commonest features are fatigue and general malaise. Most patients are asymptomatic: some patients present with established cirrhosis.

Complications
Hepatic
- Fulminant hepatitis (1–2%)
- CAH, CPH, cirrhosis
- Hepatoma.

Extrahepatic
- Aplastic anaemia, agranulocytosis
- Cryoglobulinaemia.

HCV is one of the most common causes of chronic hepatitis. For the majority, disease progression is indolent and subclinical. HCV has been implicated in type II cryoglobulinaemia, which is a vasculitis.

Diagnosis
- Non-specific features supporting HCV include:
  - consistent exposure history
  - mild hepatitis (transaminases 10×normal)
  - prolonged prothrombin time
  - relapsing course
  - negative tests for HAV IgM antibody and HBV surface antigen
- Confirmation of hepatitis C is by detection of:
  - specific antibodies to HCV by enzyme-linked immunoadsorbent assay (ELISA). Newer tests have the advantage of fewer false-positive reactions and earlier positivity following infection
  - nucleic acid (HCV RNA) by amplification (PCR). Recently infected persons will be

HCV RNA positive before they are HCV antibody positive
- Confirmation of active hepatitis C is by:
  - detection of HCV RNA. This indicates active infection; 15% of patients are PCR negative by qualitative assay and have suffered a self-limiting illness with no evidence of chronicity. Quantitative PCR is useful in predicting likelihood of response and early effects of treatment
  - liver biopsy. This should be done in all patients with a positive HCV RNA as some patients with normal LFTs have active liver disease or established cirrhosis. As for HBV, the systems for scoring severity are used (Metavir, Ishak or Knodell). This allows objective treatment decisions to be made.

Treatment
- For acute hepatitis, bed rest is necessary. In fulminant cases, intensive care is needed. Prompt treatment with interferon may prevent the development of chronic disease
- For chronic hepatitis, weekly pegylated α-interferon with daily ribavirin is indicated for all patients with moderate to severe disease (as judged by histology grading). For genotypes 2 and 3, 6 months' treatment is given with a sustained response rate (cure) of over 80% in adherent patients: for genotype 1, the success rate is 45% after 12 months' treatment. Pegylated interferon superseded standard interferon because of ease of administration and better response rates: ribavirin dosage must be weight determined to optimize success. Histological improvement parallels virological response
- Liver transplantation may be successful in carefully selected patients
- Co-infection with HIV reduces treatment response rates.

Prevention
- Elimination of high-risk persons as blood donors
- Screening of blood donors for HCV antibody
- Inactivation of HCV in blood products
- Use of synthetic plasma products produced

through recombinant DNA technology (e.g. factorVIII)
- No vaccine is available
- All patients should be immunized against HBV and HAV if not already immune on screening tests
- Low-dose interferon has been shown to reduce the incidence of hepatoma in patients with cirrhosis.

Prognosis
Up to 20% of patients with chronic hepatitis go on to develop cirrhosis within 5–30 years; 15% of these will develop hepatoma. Life expectancy for decompensated cirrhosis is less than 5 years and 1–3% will develop hepatoma annually.

Hepatitis delta
Epidemiology
HDV is a defective RNA virus which can only replicate in cells already parasitized by HBV. It can therefore only present in a person co-infected with hepatitis B. Like HBV, HDV can cause acute and chronic liver disease. This may result either from simultaneous coinfection with HBV and HDV, or from HDV superinfection in an HBV carrier. HDV:
- was originally described in Italy and is endemic in southern Europe, the Middle East, the Pacific islands and parts of Africa and South America. Transmission in this context is predominantly sexual
- has essentially been restricted to IVDUs and multiply transfused patients in Western Europe and the USA; transmission is mainly through blood products. Infection is uncommon in other groups where HBV carriage is common
- rarely results in vertical transmission, being more commonly associated with HBV e antibody positive status
- in the UK accounts for <1% of all cases of acute hepatitis, increasing to 5% where there is a history of IVDU. In chronic HBV carriers, it is found in 1–2% of non-IVDU and one-third of IVDU patients. It is the cause of 6% of all cases of chronic hepatitis, increasing to 25% where there is a history of IVDU

- where contracted simultaneously with HBV and causing acute hepatitis, is identical to acute HBV hepatitis although fulminant hepatitis is more common; many cases are subclinical, as they are with HBV. Where acute hepatitis follows on from HDV superinfection on HBV carriage, both fulminant and chronic disease is more frequent
- in chronic carriers is nearly always associated with chronic liver disease
- can be eradicated if HBV can be eliminated
- has an IP of 1–2 months.

Pathology and pathogenesis
Chronic carriage of HDV is associated with more severe liver damage on liver biopsy than HBV; rapidly progressive CAH or cirrhosis is observed. It is probable that like HBV (and unlike HAV and HCV), the virus is not directly cytopathic.

Clinical features
In acute coinfection with HDV and HBV, clinical features are no different from HBV alone (see p. 148).

Complications
- Fulminant hepatitis
- CAH, cirrhosis.
Hepatoma is not closely linked with HDV.

Diagnosis
Non-specific features supporting acute HDV include HBV surface antigen positivity, and:
- very raised transaminases (e.g. ALT or AST > 1000 IU/L)
- prolonged prothrombin time
- history of IVDU
- arrival from an endemic area
- knowledge of previous HBV carriage
- negative tests for HAV and HCV
- relapsing course and progression to chronic disease.
Confirmation of HDV is by detection of:
- HDV antibody (seroconversion may take place months after the clinical illness, so prolonged testing is necessary)
- HDV antigen

- HDV RNA by amplification (PCR)
- HDV in liver biopsy.

Treatment
Acute hepatitis
- Bed rest
- In fulminant cases, intensive care is necessary.

Chronic hepatitis
Interferon 9 megaunits thrice weekly for up to 1 year; half of the patients respond with a return of transaminases to normal and clearance of HDV RNA from serum, although many patients relapse. Rarely, HDV may be cured if HBV is eliminated. Lamivudine has no effect. Liver transplantation is an option in end-stage liver disease when liver reinfection is less likely than with hepatitis B.

Prevention
- Exclusion of high-risk persons from blood donation
- Screening of blood and blood products for HBV
- Inactivation of HDV in blood products
- Use of synthetic plasma products produced through recombinant DNA technology (e.g. factor VIII)
- Hepatitis B immunization of those at risk
- No vaccine is available.

Prognosis
Chronic HDV infection is responsible for over 1000 deaths each year in the USA and patients are significantly more likely to require liver transplantation than are patients with HBV alone.

Hepatitis E
Epidemiology
HEV is a calicivirus and epidemiologically resembles hepatitis A. Sporadic cases have been reported from all countries, but epidemic disease is mainly seen in developing nations. The first major outbreak due to HEV was in New Delhi in 1957, when there were 29 000 cases.
- Outbreaks have predominantly affected south-east Asia, Burma, Nepal, the former USSR, Mexico, Venezuela and north Africa. They are associated with gross contamination of water supplies, usually by sewage
- Secondary cases amongst household contacts do occur but less frequently than with HAV
- Transmission is by water and the faecal–oral route
- Both children and adults can be infected
- Approximately one-third of the cases of NANB hepatitis and half of those with non-A, non-B, non-C hepatitis are caused by HEV
- The seroprevalence in Europe and the USA is 1–2.5%, compared to 10–15% in south-east Asia
- There is an unusually high mortality from HEV during pregnancy (20–40%)
- The IP is 6 weeks (range 2–9 weeks).

Pathology and pathogenesis
Through molecular cloning, three HEV proteins have been expressed and form the basis of immunoassays. This will allow better understanding of the virus.

Clinical features
These are indistinguishable from HAV (see p. 145).

Complications
Fulminant hepatitis probably occurs with the same frequency (0.1%) and prognosis (20% mortality) as in hepatitis A. The incidence and mortality is particularly high in pregnant women.

Diagnosis
- Non-specific features supporting a diagnosis of hepatitis E include:
    - recent arrival from or travel to an endemic area
    - negative tests for HAV IgM antibody, HBV surface antigen and HCV antibody
    - no risk factors for hepatitis B or C
    - no evidence of chronicity, with LFTs returning to normal
- Diagnosis of HEV can be confirmed by:
    - detection of IgM and IgG antibodies to HEV

- detection of virus particles in stool by immune electron microscopy
- detection of HEV DNA by amplification
- None of the HEV-specific investigations are routinely available at present.

Treatment
The management is the same as for hepatitis A (see p. 145).

Prevention
- Passive protection using immunoglobulin may have a role, but there are few data
- Travellers should be warned of the need to adequately cook food and boil water
- No vaccine is available.

Prognosis
Outside of complications in pregnancy, when fatal fulminant hepatitis is a major problem, HEV is a relatively benign infection. Illness severity increases with age.

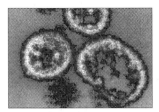

Infections of the Genitourinary Tract

Clinical syndromes

Genital ulcers

Major presenting features: *ulceration, pain, inguinal lymphadenopathy.*

Causes

- HERPES SIMPLEX VIRUS (HSV)
- *TREPONEMA PALLIDUM, HAEMOPHILUS DUCREYI*
- Lymphogranuloma venereum, granuloma inguinale (donovanosis).

Distinguishing features
See Table 11.1.

Complications
Primary syphilis: progression to secondary, latent, tertiary and quaternary disease.
Herpes simplex: recurrences, aseptic meningitis, sacral radiculomyelitis, urinary retention, secondary bacterial infection, transverse myelitis, herpes simplex infection in the 3rd trimester of pregnancy may result in neonatal herpes (see p. 99).
Haemophilus ducreyi: phimosis, inguinal buboes, auto-inoculation.

Investigations
- Ulcer swab:
  - viral culture

- dark-ground microscopy (if the lesion is non-tender)
- syphilis PCR if clinically suspected
- bacterial swab if secondary bacterial infection is suspected
- Serology:
  - for syphilis: electroimmunoassay (EIA) (IgM and IgG) if tests are positive for rapid plasma reagen (RPR) or Venereal Diseases Research Laboratory (VDRL) tests and *T. pallidum* haemagglutination assay (TPHA)
  - for *Chlamydia trachomatis* and lymphogranulovenereum (LGV) serotypes
  - if the aetiology is unclear and the ulcer persists to consider a biopsy.

Differential diagnoses
Trauma, Behçet's syndrome, intraepithelial neoplasia/invasive neoplasia, Stevens–Johnson syndrome, fixed drug eruption, skin conditions such as contact dermatitis or psoriasis, lichen sclerosis.

Treatment
- Screening for other sexually transmitted infections (STIs), partner notification and treatment if indicated
- Primary syphilis: procaine pencillin or tetracycline if pencillin-allergic; longer course of treatment if HIV positive
- Herpes simplex: valaciclovir/acyclovir

	Primary syphilis	Genital herpes	Haemophilus ducreyi (chancroid)
Incubation period	9–90 days	2–3 weeks	3–14 days
Ulcer number	Single	Multiple	Single/multiple
Edge	Erythematous, well-defined, rounded	Erythematous, small, round	Erythematous, ragged, undermined
Induration	Present	Absent	Absent
Pain	Absent	Marked	Marked
Tenderness	Absent	Marked	Marked
Base	Clean	Clean	Dirty
Slough	Clear	Grey-white	Necrotic
Size where multiple	Uniform	Uniform	Variable
Lymph-node swelling	Moderate	Moderate and tender	Marked
Tenderness	Nil	Marked	Marked
Systemic symptoms	Absent	If primary infection	Absent
Diagnosis	Dark-ground microscopy, serology, direct fluorescent antibody test, PCR*	Culture	Gram stain: very insensitive and not very specific. Culture: the media is not widely available. PCR is available but is also currently being validated

* Polymerase chain reaction (PCR) is a new test at the time of writing which is currently being validated.

Table 11.1 Distinguishing features of the main infective causes of genital ulcers.

• *Haemophilus ducreyi:* azithromycin, ceftriaxone, ciprofloxacin or erythromycin.

Prevention
• Herpes simplex: acyclovir can prevent recurrences
• Primary syphilis: progression can be prevented by adequate initial treatment
• Counselling as to safe sex, use of condoms, etc.

Urinary tract infection and urethritis
Major presenting features: *dysuria, frequency, urethral discharge.*

Urinary tract infection
Causes
Predisposing
Lower urinary tract infection (UTI):
• obstruction (prostatic hypertrophy, urethral valves or stricture)
• poor bladder emptying (neuropathic, diverticula)
• catheterization/instrumentation
• vesicoenteric fistula
Upper UTI:
• vesicoureteric reflux
• obstruction (e.g. calculus, stricture).

Microbiological
• *Escherichia coli, Proteus* spp., *Klebsiella* spp., other 'coliforms', *Enterococcus faecalis, Staphylococcus aureus, S. saprophyticus, S. epidermidis, Pseudomonas aeruginosa*
• *Mycobacterium tuberculosis, Schistosoma haematobium,* brucellosis.

Complications

Lower UTI: pyelonephritis, septicaemia, epidymitis, prostatitis, chronic cystitis.

Upper UTI: perinephric abscess, chronic pyelonephritis and scarring, septicaemia, renal failure, renal stones, ureteric stricture.

Investigations

All
• Midstream urine (MSU) for microscopy, culture and sensitivity.

Upper UTI
• Blood culture
• Full blood count (FBCt), differential white cell count (WCCt) and erythrocyte sedimentation rate (ESR)
• Biochemical profile (liver function tests (LFTs), albumin and urea/creatinine)
• Chest X-ray (CXR) (to exclude pneumonia).

Differential diagnoses

Upper UTI
Lower lobe pneumonia, cholecystitis, muscoskeletal pain, perinephric abscess, intraperitoneal abscess, pelvic inflammatory disease (PID).

Treatment

Upper UTI
• Intravenous (IV) antibiotics (e.g. cefuroxime).

Lower UTI
• Oral antibiotics, e.g. one of trimethoprim, co-amoxiclav or ciprofloxacin.

Urethritis (in men)

Symptoms
Urethral discharge, dysuria and urethral discomfort.

Aetiology
• *Neisseria gonorrhoeae*
• Non-gonococcal urethritis (NGU)
    • *Chlamydia* accounts for 30–50% of cases
    • The aetiology of most cases of non-chlamydial NGU is unknown
    • *Ureaplasma urealyticum*, *Mycoplasma gentilium*, *Trichomonas vaginalis*, HSV, *Candida*, *Neisseria meningitidis*, urethral stricture and foreign bodies account for a small proportion of cases.

Distinguishing features
See Table 11.2.

Complications
Epididymo-orchitis, Reiter's syndrome and transmission to the partner.

Investigations
• Urethral swab: for Gram stain and culture on selected media for gonococcus
• First-catch urine for *Chlamydia* polymerase chain reaction (PCR).

Treatment
• Screening for other STIs, partner notification and treatment (if indicated)

Table 11.2 Distinguishing features of gonococcal and non-gonococcal urethritis.

	Gonococcal urethritis	Non-gonococcal urethritis
Incubation period	< 1 week	> 1 week
Onset	Abrupt	Insidious
Symptoms	Constant, severe	Intermittent, may remit
Dysuria and discharge	Sometimes	50% (approximate)
Discharge	Purulent	Mucoid
Gram stain	Positive for intracellular diplococci	Positive for polymorphs (> 5 per high-power field)
	Lots of polymorphs	

- *Neisseria gonorrhoeae:* ciprofloxacin, or cephalosporin such as ceftriaxone
- *Chlamydia trachomatis:* azithromycin or doxycycline
- If the urethritis is not responsive to any of those then a 2-week course of erythromycin and metronidazole should be prescribed
- A test of cure should be performed for gonorrhoea.

Vaginal discharge

Most often the result of one or more of these three common infections.
- Bacterial vaginosis (the commonest cause of vaginal discharge in women of childbearing age). It is due to a depletion of the lactobacilli in the vagina leading to an increase in vaginal pH and overgrowth of anaerobic and other bacteria. It is not sexually transmitted and treating partners is of no benefit
- Candidiasis—usually due to *Candida albicans* but may be due to other species, most commonly *glabrata* and *tropicalis*. It affects 75% of women at some point in their lives and 10–20% of women are asymptomatic carriers of *Candida*
- *Trichomonas vaginalis*—a sexually transmitted infection due to a flagellate protozoon.

Table 11.3 Distinguishing features of vaginal discharge.

Distinguishing features

See Table 11.3.

Complications and treatment

Bacterial vaginosis
- Miscarriage, preterm birth, postpartum or post-termination PID are associated risks
- Treat with metronidazole if the woman is symptomatic, pregnant or about to undergo gynaecological surgery.

Trichomonas vaginalis

- Symptomatic women and their sex partners receive metronidazole
- Though an association with adverse pregnancy outcomes has been reported, treating asymptomatic pregnant women does not lessen that risk.

Candidiasis

- See p. 196 and 197.

Pelvic inflammatory disease (PID)

Major presenting features: *lower abdominal pain, vaginal discharge, fever.*

Causes

Ascending infection from endocervix. *N. gonorrhoeae* and *Chlamydia trachomatis* have been shown to be causal. Anaerobic bacteria are also associated with PID.

	Bacterial vaginosis	Trichomonas vaginalis	Candidiasis
Discharge	Homogenous Fishy odour	Homogenous May be malodorous	Curdy Not malodorous
Vulvovaginitis	No	Yes	Yes
Cervicitis	No	Maybe 'strawberry cervix': erythematous seen in about 2% of people	Yes/No
Asymptomatic	50%	10–50%	10–20%
Diagnosis	Presence of 3 or more of Amsel's criteria (homogenous discharge, pH > 4.5, clue cells on microscopy, strong odour on mixing KOH with the discharge)	Observing the motile trichomonads in wet-mount microscopy. Culture	Culture. Wet mount or Gram-staining microscopy

Predisposing
• Young age, multiple sexual partners, recent new partners, use of non-barrier or no contraception, termination of pregnancy.

Microbiological
• *Neisseria gonorrhoeae*, *Chlamydia trachomatis*
• *Escherichia coli*, other 'colifoms', *Enterococcus faecalis*, anaerobes
• *Mycobacterium tuberculosis*, actinomycosis.

Clinical features
• Lower abdominal pain
• Dyspareunia
• Vaginal discharge—abnormal
• Irregular *per vaginum* bleeding
• Lower abdominal tenderness
• Cervix motion tenderness
• Adnexal tenderness
• Fever.

Differential diagnosis
• Ectopic pregnancy
• Endometriosis
• Appendicitis
• Ovarian cyst accident.

Distinguishing features
See Table 11.4. Clinically large overlap observed.

Complications
Infertility, ectopic pregnancy, chronic pelvic pain, perihepatitis.

Table 11.4 Distinguishing features of pelvic inflammatory disease.

Investigations
• Endocervical swabs: *Chlamydia* PCR, gonococcal culture
• High vaginal swab: culture
• Urine: *Chlamydia* PCR
• Ultrasound if abscess suspected
• If the diagnosis is uncertain laparoscopy may be required.

Treatment
• Ofloxacin and metronidazole
• Erythromycin plus metronidazole if pregnancy suspected
• If severe systemic symptoms IV cephalosporin plus IV then oral doxycyline plus IV metronidazole, or IV ofloxacin plus IV metronidazole
• Screening for other STIs, partner notification and treatment if indicated.

Specific infections

Gonorrhoea
Epidemiology
Gonorrhoea is a mucosal infection of the columnar epithelium transmitted by sexual intimacy and caused by *Neisseria gonorrhoeae*.
• Occurs worldwide, higher prevalence in many developing countries
• Attack rates are highest in those aged 15–24 years who live in cities, belong to a low socioecomic group, are unmarried or homosexual, or have a past history of sexually transmitted disease (STD)
• The disease is highly transmissible, with infection rates of 50% in women and 20% in men after a single unprotected vaginal exposure

	Gonococcal PID	Non-gonococcal PID
Onset	Abrupt	Insidious
Fever	≥ 38°C	≤ 38°C
Toxicity	May be marked	Mild
Endocervical discharge	Purulent	Mucoid or mucopurulent
Abdominal pain/guarding	Usually marked	Less marked
White cell count ($\times 10^9$/L)	≥ 20	< 20

• Approximately 75% of women are asymptomatic, compared to only 5% of heterosexual men
• Extragenital sites of infection include the oropharynx, eyes, and perihepatic tissues; disseminated infection is rare
• The incidence steadily increased between 1951 and 1980, after which it fell but in recent years started to rise again, particularly in homosexual men; there are approximately 12000 cases in England per annum.
• Severe systemic infections and ophthalmia neonatorum have become rare in developed nations
• Protective immunity does not develop and re-infections are common after re-exposure
• The incubation period (IP) is 2–7 days.

Pathology and pathogenesis
Following attachment, gonococci penetrate into the epithelial cells and pass to the subepithelial tissues, where they are exposed to the immune system (serum, complement, immunoglobulin A (IgA), etc.) and phagocytosed by neutrophils. Virulence is dependent upon gonococci being able to attach to and penetrate into host cells as well as being resistant to serum, phagocytosis and intracellular killing by polymorphonucleocytes. Factors conferring these virulence factors include pili, the outer membrane proteins (OMPs), lipopolysaccharide and IgA proteases.

Clinical features
In men
• Urethritis is associated with a purulent urethral discharge, dysuria and a hazy urine on first micturition
• Anorectal gonorrhoea is associated with:
  • perianal pain, pruritus, a mucoid or mucopurulent discharge or anal bleeding: infection is asymptomatic in 60%
  • a distal visible (20%) and histological (40%) proctitis.

In women
Infection is associated with:
• vaginal discharge, dysuria, frequency; in many, examination is normal

• cervicitis, urethritis or proctitis in decreasing order of frequency. Cervicitis is characterized by an erythmatous friable cervix and mucopurulent discharge.

Complications
In both sexes
• Dissemination
• Perihepatitis (Fitz-Hugh–Curtis syndrome)
• Endocarditis, meningitis (both rare).
Dissemination occurs in 1–2% and manifests with a large-joint arthritis or tenosynovitis, a sparse maculopapular/pustular rash, usually on the extremities, and systemic symptoms. It is more common in women.

In men
• Prostatitis, epididymitis
• Urethral stricture, periurethral abscess or penile lymphangitis.

In women
• PID, sterility
• Bartholinitis/abscess formation
• Neonatal conjunctivitis in offspring.

Diagnosis
Diagnosis is by identification of the organism, either by microscopy of Gram-stained smears or culture (sensitive and specific).
• Gram staining of urethral discharge is positive in 95% of men and of endocervical discharge in 60% of women
• Culture is essential in women, including rectal and oropharyngeal cultures — 10% is extragenital. Confirmation of identity can then be made by sugar fermentation or *N. gonorrhoeae*-specific antigen detection kits.

Nucleic acid hybridization or amplification tests are useful non-culture tests for screening.

Treatment
Penicillin resistance may be mediated by plasmid-encoded penicillinase or chromosomally encoded changes to penicillin-binding proteins. It accounts for 3–5% in the UK.
• Fluoroquinolone-resistant strains are becoming common in parts of Asia (40%)

- Because of the importance of eradication to prevent further transmission, single-dose therapy is preferable, with either:
  oral ciprofloxacin or
  IM ceftriaxone which should be used for patients who fail to respond to the above or oral amoxycillin (high dose 3 g) in areas of low-level penicillin resistance, or in pregnancy
- Sexual contacts must be examined, screened and treated if necessary

Syphilis
Epidemiology
Syphilis is a chronic, systemic, infectious disease characterized by clinically well-defined stages. It is caused by *Treponema pallidum* which belongs to a family of spirochaetes (*Borrelia, Leptospira, Spirillum* and *Treponema*).
- *Treponema pallidum* cannot be cultured. Treponemal cardiolipin cross-reacts with host-tissue antigen which is the basis of the use of VDRL in diagnosis
- Incidence had been steadily declining (1000 cases/year in the UK in the mid-1990s), but in recent years there has been a resurgence in early, infectious syphilis in the UK. Congenital syphilis is exceptionally rare in developed nations. However, congenital and acquired disease remain important problems in developing countries
- Disease may be congenital or acquired, each being subdivided into early (duration <2 years from infection) and late (duration >2 years) stages. Early acquired syphilis is further subdivided into primary, secondary and early latent stages
- Late acquired syphilis is split into late latent, tertiary (reserved for benign gummatous syphilis) and quaternary (including all other manifestion of late syphilis)
- Patients may progress through stages sequentially, or some clinical stages may be clinically inapparent. Approximately 40% of patients with untreated latent syphilis go on to develop late-stage disease
- Transmission is by sexual contact, occasionally by intimate non-sexual contact, rarely through inoculation injury, or via blood transfusion and also by vertical transmission *in utero*

- Non-venerally acquired treponemas are phenotypically indistinguishable and include *T. pertenue* (yaws), *T. carateum* (pinta), and *T. pallidum* var. endemic syphilis (bejel). They do not affect the central nervous system and are geographically localized, whereas syphilis is worldwide
- Syphilis complicating human immunodeficiency virus (HIV) infection may be accelerated and atypical
- The IP is 17–28 days (range 9–90 days).

Pathology and pathogenesis
Obliterative endarteritis occurs at all stages of the disease. It is associated with perivascular infiltration with macrophages and plasma cells in the primary chancre, hyperkeratosis in cutaneous secondary syphilis, and central necrosis and granulomata in gummas. Treponemas enter the body through mucous membranes or abraded skin where the primary chancre develops. This heals spontaneously over 2–4 weeks and its disappearance coincides with the secondary or disseminated stage, when treponemes can be identified from the skin lesions as well as from blood, lymph nodes and the central nervous system.

Clinical features
The primary chancre in 95% of cases is genital and has the following characteristics.
- It is a usually solitary, painless, non-tender, rounded lesion with a well-defined erythematous margin and an indurated clean base
- It is associated with rubbery, open, painless, non-tender inguinal lymphadenopathy
- It is an ulcer that heals without scarring by 4–6 weeks.

Complications
Secondary syphilis: 6–8 weeks from infection
The signs of secondary syphilis are numerous and may be preceded by mild constitutional symptoms of fever and influenza-like symptoms. Characteristic features are:
- a maculopapular rash (90% of cases) which is non-pruritic and symmetrical and involves the entire body including the face, scalp, palms and soles. In the perianal area this is manifested as

condylomata lata, in the mouth as mucosal ulcers (snail-track or mucous patches) and in the hair as patchy alopecia
• generalized and painless moderate lymphadenopathy (50%), especially of the epitrochlear glands
• uveitis, hepatatis and glomerulonephritis in a few patients.

Tertiary syphilis
Tertiary syphilis (skin, mucous membranes and bones) may develop 3 or more years after the primary stage and the characteristic lesion of this stage is the gumma.
• A cutaneous gumma is usually a punched-out ulcer with a 'wash-leather' base, central depigmentation and peripheral hyperpigmentation which heals to form a 'tissue-paper' scar
• Destructive mucosal gummas, periostitis (sabre tibia), gummas of the liver and uveitis may also occur.

Quaternary syphilis
Quaternary syphilis has its major effects on the heart and nervous system.

Cardiovascular syphilis
• Occurs in approximately 10% of untreated patients (twice as many men as women)
• It is primarily an aortitis with secondary aortic incompetence, coronary ostial stenosis (leading to angina) and eventually aneurysmal dilatation of the aorta
• It usually affects the ascending aorta ('aneurysm of signs' — sternal/rib erosion, right heart failure, lung collapse) or the arch ('aneurysm of symptoms' — hoarseness, brassy cough, stridor, dysphagia).

Neurosyphilis Neurosyphilis is divided into meningovascular, parenchymatous and tabes dorsalis, although there is a great deal of overlap.
• It occurs in approximately 20% of untreated patients
• Meningovascular syphilis may appear as early as 5 years after initial infection and presents with an aseptic meningitis, cranial nerve palsies and occasionally hemiplegia

• Generalized paresis of the insane occurs later, at 10–30 years, with global cortical dysfunction leading to cognitive impairment, loss of inhibition, tremors, fits, Argyll Robertson pupils and, eventually, dementia
• Tabes dorsalis occurs at 15–35 years after primary syphilis. It is characterized by lightning pains, loss of sensory modalities (postural, temperature, deep and superficial pain), hypotonia, ataxia, Charcot's joints, neuropathic ulcers, loss of tendon reflexes and bladder disturbances.

Congenital syphilis
Congenital syphilis may result from infection *in utero* and manifest early, with snuffles, maculopapular rash, osteochondritis (saddle nose, sabre shin), hepatosplenomegaly and anaemia; or late, with interstitial keratitis, frontal bossing, deafness, abnormal dentition and recurrent arthropathy.

Diagnosis
Primary and secondary syphilis
• Dark-ground illumination of a wet-mount preparation taken from the chancre, condylomata or mucous lesion, showing characteristic spirochaetes. PCR test and direct fluorescent antibody test are all useful
• Non-specific (VDRL/RPR — detecting cardiolipin antibody) and specific (TPHA, fluorescent treponemal antibody (FTA) — detecting treponemal antibody) serological assays. In primary syphilis, the FTA precedes the TPHA, which in turn precedes the VDRL in becoming positive. In secondary syphilis, all three tests are strongly positive. The VDRL titre is an indicator of disease activity; specific tests remain positive for life. False-positive non-specific tests are not infrequent and can occur transiently in many infections, as well as in systemic lupus erythematosus.

Latent, tertiary and quaternary syphilis
Latent syphilis is diagnosed by serology, tertiary and quaternary syphilis by a combination of clinical features and serology.
• Up to 25% of patients with late syphilis have

negative non-specific tests, although all have specific antibodies
• Histopathology is diagnostic but this is rarely obtainable, except in skin and mucous membrane lesions
• In a patient with positive serology:
  • calcification of a dilated ascending aorta with aortic incompetence is very suggestive of syphilitic aortitis
  • an abnormal cerebrospinal fluid (raised protein, > 5 white cells/µL, raised globulin) is suggestive of neurosyphilis
• Distinction of syphilis from non-venereal treponemal infection is impossible by serology.

Treatment
Primary and secondary syphilis
• Either procaine penicillin daily for 14 days, a single dose of benzathine penicillin G, or tetracycline or erythromycin for 14 days (if penicillin-allergic)
• Prolonged course required if HIV co-infected — treat as for neurosyphilis.

Tertiary or quaternary syphilis
• Tetracyclines (doxycycline) for 28 days
• Procaine penicillin and oral probenecid for 17 days
• Corticosteroids are sometimes indicated in late-stage syphilis or in pregnancy to reduce the risk of a Herxheimer reaction.

Prevention
Up to half of the sexual contacts are infected; contact tracing is vital.

Genital chlamydia
Genital infection by the bacterium *Chlamydia trachomatis* is the commonest STI in the industrialized countries.

Epidemiology
• Between 0.5 and 1 million cases probably occur annually in the UK (3 million in the USA); the vast majority are asymptomatic and remain undiagnosed
• Most common in women aged 16–24 and men aged 20–24 years

• Infection is transmitted through unprotected vaginal, anal or oral sex with an infected partner
• Multiple partners increase risk
• Infected persons are frequently asymptomatic but still infectious
• Pregnant women can pass infection to infants during birth
• Infection can be transmitted to the eyes through contaminated fingers
• The IP is 1–3 weeks in those developing symptoms.

Clinical features
• Up to 70% of infected women and 50% of men are asymptomatic
• About 40% of women will develop PID (chlamydia accounts for 40% of all PID cases)
• Men may develop urethritis; proctitis or proctocolitis (in those engaging in receptive anal intercourse)
• Conjunctivitis
• Conjunctivitis and pneumonia in the newborn.

Complications
• The consequences of PID are often severe, regardless of symptom severity:
  • 20% will become infertile
  • 18% will have chronic pelvic pain
  • 9% will have a life-threatening ectopic pregnancy
• Epididymitis in men (rare)
• Reactive arthritis (more common in men than women).

Diagnosis
• PCR of genital swab (vaginal, cervical or anal) or urine
• National screening programmes are necessary to diagnose the large number of asymptomatic cases (initiated in the USA and recommended in the UK).

Treatment
• Doxycycline for 7 days or single-dose azithromycin
• Testing and treatment of all current and recent sexual partners.

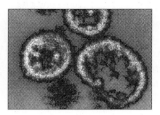

Bone and Joint Infections

Clinical syndromes

Osteomyelitis and infective arthritis

Major presenting features: *joint or bone pain, tenderness, redness with or without swelling, fever.*

Causes
Osteomyelitis
Predisposing
• Compound fracture, dental infections (jaw), postcardiac surgery (sternal), soft tissue infection (pressure sores — sacrum)
• Sickle cell disease, metal implant, diabetes mellitus (foot), peripheral vascular disease (foot).

Microbiological
• STAPHYLOCOCCUS AUREUS, *Streptococcus pyogenes*, *Haemophilus influenzae* (type b), *Salmonella typhi*, other *Salmonella*, *Escherichia coli*, other 'coliforms', *Pseudomonas aeruginosa*, anaerobes
• MYCOBACTERIUM TUBERCULOSIS, actinomycosis, hydatid disease.

Infective arthritis (native joint)
Predisposing
• Pre-existing arthritis (especially rheumatoid), intra-articular injection, metal implant, trauma, neighbouring osteomyelitis, corticosteroids, malignancy, intravenous (IV) drug abuse.

Microbiological
• STAPHYLOCOCCUS AUREUS, *S. pyogenes*, group B β-haemolytic streptococci, *H. influenzae* (type b), *Neisseria gonorrhoeae*, *E. coli*, other 'coliforms', *Salmonella*, *Neisseria meningitidis*, *P. aeruginosa*
• PARVOVIRUS, MYCOPLASMA PNEUMONIAE, rubella, hepatitis B, mumps, Epstein–Barr virus, adenovirus, influenza, arboviruses
• Lyme disease, rat-bite fever
• MYCOBACTERIUM TUBERCULOSIS, brucellosis
• Reactive arthritis (enteric pathogens, *Chlamydia trachomatis*), infective endocarditis, rheumatic fever (*S. pyogenes*).

Infective arthritis (prosthetic joint)
Predisposing
• Rheumatoid arthritis, corticosteroids, diabetes mellitus, obesity, advanced age, prior surgery
• Wound haematoma or infection.

Microbiological
• STAPHYLOCOCCUS EPIDERMIDIS, S. AUREUS, Gram-negative bacilli, anaerobes, other streptococci, diphtheroids, *Candida*.

Distinguishing features
See Table 12.1.

Complications
Osteomyelitis: chronic osteomyelitis (involucrum/sequestrum formation), septic arthritis, septicaemia, amyloidosis, sinus formation.
Septic arthritis: reduced joint movement, destructive arthritis, osteoarthritis, osteo-

	Septic arthritis	Viral arthritis
Onset	Usually acute	Usually gradual
Fever	Moderate to marked	Nil to moderate
Toxicity	Moderate to marked	Minimal
Number of joints	Single	Usually multiple
Pain, redness, swelling	Marked	Mild to moderate
Blood culture	Positive in 50%	Negative
Joint aspirate	Turbid/purulent	Clear, straw coloured
White cell count	> 100 000/mm³, neutrophilic	< 100 000/mm³, lymphocytic
Gram stain	Often positive	Negative
Bacterial culture	Positive	Negative
Peripheral white cell count differential	Raised, neutrophilic	Normal/low, lymphocytic
Diagnosis	Gram, culture	Serology

Table 12.1 Distinguishing features of infective arthritis.

myelitis, contractures, septicaemia, sinus formation, avascular necrosis (hip).

Prosthesis infection: loosening, sinus formation, low-grade bacteraemia, chronic infection unless prosthesis removed.

Investigations

Osteomyelitis and arthritis
• Full blood count, differential white cell count, erythrocyte sedimentation rate
• Biochemical profile (alkaline phosphatase, calcium, phosphate, liver function tests, urea and creatinine)
• Blood culture
• Midstream urine (may be the source in elderly)
• X-rays of affected joint/bone (for lytic lesions, periosteal reaction: takes 2 weeks for abnormalities to develop).

Arthritis
• Autoimmune profile (rheumatoid factor, antinuclear factor, double-stranded DNA), urate; rheumatoid arthritis, systemic lupus erythematosus (SLE) and gout may mimic septic arthritis

• Ultrasound scan (especially for hip) to demonstrate any effusion
• Joint aspirate:
  • cell count, Gram stain, Ziehl–Neelsen stain
  • culture
  • crystals (to exclude gout, pseudogout)
• Serology:
  • viral (mumps, rubella, parvovirus, influenza)
  • antistreptolysin O titre
  • *Mycoplasma pneumoniae*
• Arthrography (if chronic)
• Synovial biopsy (where negative cultures in chronic disease).

Osteomyelitis
• Isotope bone scan (technetium)
• Computed tomography or magnetic resonance scan (benefit in identifying involucrum/sequestrum)
• Bone biopsy (as above).

Differential diagnoses

Osteomyelitis: osteosarcoma, benign bone tumour, sickle cell thrombotic crisis.

Septic arthritis (native joint): rheumatoid arthritis, SLE, Still's disease, seronegative arthritis (Reiter's syndrome, psoriatic, inflammatory bowel disease), gout, pseudogout, rheumatic fever, sarcoidosis, bursitis, haemarthrosis, irritable hip.

Septic arthritis (prosthetic joint): mechanical loosening.

Treatment

Acute septic arthritis or osteomyelitis
- Antibiotics: initially IV (2 weeks) followed by oral (2–6 weeks):
  - *S. aureus*: flucloxacillin and initially either fucidin or rifampicin or gentamicin
  - unknown: flucloxacillin and cefotaxime — alter as appropriate when identification and sensitivities available
- Aspiration (joint) or curettage (bone). Rarely, open surgical drainage of a joint is necessary

- Bed rest and immobilization until joint/bone inflammation has subsided
- Then intensive physiotherapy.

Infected prosthetic joint
- Removal of prosthesis
- Antibiotics: initially IV (2–4 weeks) followed by IV or oral:
  - *S. epidermidis*: vancomycin (or teicoplanin) with or without another antibiotic to which the organism is sensitive.

Prevention
Antibiotic prophylaxis for implant surgery.

Multisystem Infections

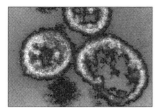

Acquired Immunodeficiency Syndrome

Introduction

• The acquired immunodeficiency syndrome (AIDS) was first recognized in 1981 and is caused by the human immunodeficiency virus (HIV-1)
• In the subsequent two decades, AIDS has grown to be the second leading cause of disease burden worldwide and the leading cause of death in Africa
• The virus is an RNA retrovirus from the lentivirus family
• HIV-2 causes a similar illness to HIV-1, although less aggressive, and is restricted mainly to western Africa
• The viruses almost certainly originated from closely related African primate viruses
• Immune deficiency is a consequence of continuous high-level HIV replication leading to

virus- and immune-mediated destruction of the CD4+ helper T lymphocytes. These cells are responsible for the initiation of the majority of immunological responses to pathogens
• Following infection by HIV there is gradual loss of CD4 cells, resulting in increasing impairment of cell-mediated immunity with consequent susceptibility to opportunistic infections (OIs) and HIV-related tumours.

Epidemiology

Transmission

• HIV is present in blood, semen and some other body fluids (e.g. breast milk and saliva)
• After exposure to infected fluid the risk of contracting infection is dependent on the viral load, the integrity of the exposed site, and the type and volume of body fluid

• Transmission can be sexual, parenteral (blood or blood product recipients, injection drug users and occupational injury) or vertical
• Transmission risk after a single exposure is >90% for blood or blood products, 14% for vertical, 0.5–1.0% for injection drug use, 0.2–0.5% for genital mucous membrane and <0.1% for non-genital mucous membrane
• Mother-to-child transmission (MTCT) is higher (up to 40%) in developing countries. Zidovudine (ZDV) alone (reduction to 7%), in combination with elective Caesarean (2%), or highly active antiretroviral therapy (HAART) (<1% if viral load <50 copies/mL) can decrease the risk of transmission. MTCT has now been reduced by over 90% in the developed world. This emphasizes the importance of prenatal screening of mothers for HIV infection
• 80% of those infected vertically are infected close to the time of delivery
• 70% of patients with haemophilia A and 30% of those with haemophilia B had been infected through contaminated blood products by the time HIV antibody screening was adopted in the USA and Europe (1985)
• The risk of HIV transmission with a single blood unit is now $1/10^6$ and represents blood donors in the seroconverting phase of infection
• There have been approximately 100 definite and 200 possible cases of HIV acquired occupationally in health-care workers. The risk with needlestick injury is 0.32% and with mucous membrane exposure 0.03%.

Factors associated with increased risk of HIV transmission are listed in Table 13.1.

Global situation
• Many different cultural, social and behavioural aspects determine the characteristics of HIV disease in each region
• Seroprevalence rates among injection drug users vary enormously around the world but current epidemics are occurring in eastern Europe, Russia and northern India
• There were an estimated 42 million people living with HIV/AIDS in 2002, with over 5 million new infections and 3.0 million deaths

• In Africa, HIV now affects 25–40% of adults in Botswana, South Africa and Zimbabwe and over 10% in most other African countries excluding north Africa
• Worldwide, heterosexual transmission accounts for >85% of infections
• In the USA and northern Europe, the epidemic has mainly been in men who have sex with men, whereas in southern and eastern Europe, Vietnam, Malaysia, north-east India and China, the incidence has been greatest in injection drug users. In Africa, South America and much of south-east Asia the dominant route of transmission is heterosexual and vertical
• Heterosexual transmission is now responsible for 25–30% of new infections in Europe and the USA with racial and ethnic minorities representing an increasing fraction. In the UK, three-quarters of patients with heterosexually acquired infection have recently arrived from a country with a high prevalence of HIV, mainly sub-Saharan Africa
• High-risk sexual behaviour is also increasing, with rates of rectal gonorrhoea and primary syphilis rising in the UK, USA and the Netherlands. Some of this relates to a misperception that HIV is 'treatable'.

Pathogenesis

Virology
• The HIV virion is round with a lipid membrane lined by a matrix protein and studded with glycoprotein (gp) 120 and gp41 spikes. This surrounds a cone-shaped protein core housing two copies of the ssRNA genome and viral enzymes
• Attachment is initially between the gp120 and CD4 cell receptor, which triggers conformational change in the gp120 enabling binding to a chemokine coreceptor (usually CCR5 or CXCR4). Following this, a fusion pore develops, mediated by gp41
• Once in the CD4 cell, a DNA copy is transcribed from the RNA genome by the reverse transcriptase (RT) enzyme that is

carried by the virus. This is a very error-prone process

• This DNA is then transported into the nucleus and integrated randomly within the host cell genome. Integrated virus is known as proviral DNA

• On host cell activation, RNA is transcribed from this DNA template and subsequently translation results in the production of viral proteins

• Precursor polyproteins are cleaved by viral protease to enzymes (e.g. reverse transcriptase and protease) and structural proteins. These are then used to produce infectious viral particles, which bud from the cell surface incorporating the host cell membrane

• The new infectious virus (virion) is then available to infect uninfected cells and repeat the process

• There are three groups (nearly all infections are group M) and 10 subtypes (B predominates in Europe) for HIV-1.

Immunology

• During the course of HIV infection there is a gradual reduction in the number of circulating CD4 cells that is inversely correlated with the plasma viral load. The exact mechanisms underlying the decline are not well understood

• Because CD4 cells are pivotal in the immune response, any depletion in numbers renders the body susceptible to opportunistic infections and oncogenic virus-related tumours

• Lymphatic tissues (spleen, lymph nodes, tonsils/adenoids, etc.) serve as the main reservoir of HIV infection. The virus can infect the nervous system directly.

Clinical features

Primary infection (seroconversion)

• This is usually symptomatic (70–80%) and occurs 3–12 weeks after exposure

• Over half have fever, rash and cervical lymphadenitis. Rarely, presentation may be neurological (aseptic meningitis, encephalitis, myelitis, polyneuritis)

• It coincides with a surge in plasma HIV RNA levels to >1 million copies/mL (peak between 4 and 8 weeks), and a fall in the CD4 count to 300–400 cells/mm³ but occasionally to below 200 when opportunistic infections (e.g. oropha-

Transmission	Factors increasing risk
Common to all	High viral load Presence of AIDS Seroconversion Low CD4 count
Mother to child	Prolonged rupture of membranes Vaginal delivery Breast feeding No HIV prophylaxis
Sexual	Coexistent STDs Receptive vs. insertive anal sex Non-circumcised Increased number of partners
Injection drug use	Sharing equipment and using frequently IV vs. subcutaneous injecting
Occupational	Deep injury Visible blood on device Previous arterial or venous device siting

Table 13.1 Transmission risk factors.

ryngeal candidiasis, *Pneumocystis carinii* pneumonia, PCP) may occur
- Symptomatic recovery occurs after 1–2 weeks although the CD4 count rarely recovers to its previous value
- Diagnosis is made by detecting HIV RNA in the serum or by immunoblot assay (which shows antibodies developing to early proteins)
- The appearance of specific anti-HIV antibodies in serum (seroconversion) takes place later at 3–12 weeks (median 8 weeks) although very rarely seroconversion may take place after 3 months
- The level of the viral load post seroconversion strongly correlates with the risk of subsequent progression of disease. Other factors predicting faster HIV progression are a history of seroconversion illness, evidence of candidiasis and neurological involvement
- The differential diagnosis of primary HIV includes acute Epstein–Barr virus (EBV), cytomegalovirus (CMV), streptococcal pharyngitis, toxoplasmosis and secondary syphilis
- A high index of suspicion should be maintained if a patient presents with a Monospot-negative, mononucleosis-like syndrome, when risk factors for HIV are present.

Asymptomatic phase (category A CDC classification)
- Over a variable time, the infected individual usually remains well with no evidence of HIV disease except for the possible presence of persistent generalized lymphadenopathy (PGL; defined as enlarged glands in two or more extrainguinal sites)
- Depending on the size of the viral load, there is an inverse fall in the CD4 count by usually between 50 and 150 cells/year.

Symptomatic phase (category B)
- Clinical evidence of mild impairment of the immune system then develops in many and represents the crossing over from being clinically well to the syndromes associated with AIDS
- By definition, these conditions are not AIDS

defining and include chronic weight loss, fever or diarrhoea (but not fulfilling criteria for AIDS), oral or vaginal candidiasis, oral hairy leucoplakia (OHL), recurrent herpes zoster infections, severe pelvic inflammatory disease, bacillary angiomatosis, cervical dysplasia and idiopathic thrombocytopenic purpura.

AIDS (category C)
- Late-stage disease is present when CD4 counts fall to $<200/mm^3$ and/or AIDS has developed. The term 'AIDS' is of limited use since advances in treatment and primary prophylaxis have delayed or prevented many AIDS-defining diagnoses from developing
- Very advanced disease is associated with CD4 counts of $<50/mm^3$ and mortality is highest in this group of patients. However, previously undiagnosed patients may present at this time having been in previous good health. Most commonly, however, these patients present with a history of worsening general health and acute PCP.

Correlation between CD4 count and disease
See Table 13.2.

Diagnosis
- Confirmation is by antibody testing against structural viral antigens. False-positive and false-negative results are rare
- For vertical transmission (HIV antibody positive) and seroconversion (HIV antibody negative), serology is unhelpful and HIV RNA must be measured. This is based on nucleic acid amplification
- For monitoring disease progression, the viral load (VL) and CD4 count are measured regularly (8–12 weekly). VL pretreatment determines the speed of CD4 fall, and post treatment the success of treatment (defined as a VL of <50 copies/mL). The CD4 count determines the likelihood of complications; CD4 counts of >200 cells/mm³ represent limited risk.

> 500 cells/mm³	Persistent generalized lymphadenopathy Recurrent vaginal candidiasis
200–500 cells/mm³	Pulmonary tuberculosis *Kaposi's sarcoma* Herpes zoster HIV-associated ITP Oropharyngeal candidiasis CIN II–III Oral hairy leucoplakia Lymphoid interstitial pneumonitis *Salmonellosis**</br>
< 200 cells/mm³	*Pneumocystis carinii pneumonia* *Oesophageal candidiasis* *Mucocutaneous herpes simplex†* *Miliary/extrapulmonary tuberculosis* *Cryptosporidium†* *HIV-associated wasting* Microsporidium Peripheral neuropathy
< 100 cells/mm³	*Cerebral toxoplasmosis* *Non-Hodgkin's lymphoma* *Cryptococcal meningitis* *HIV-associated dementia* *Primary CNS lymphoma* *Progressive multifocal leucoencephalopathy*
< 50 cells/mm³	CYTOMEGALOVIRUS RETINITIS/GASTROINTESTINAL DISEASE Disseminated *Mycobacterium avium-intracellulare*

Small capitals denotes AIDS-defining diagnoses.
* Recurrent non-typhi.
† Chronic.

Table 13.2 Correlation between CD4 count and disease.

Baseline investigations that are indicated
See Table 13.3.

Site-specific complications

HIV is unique for the range of organisms and malignancies that occur. Most of these have an association with one site but in the face of profound CD4 cell loss (<50 cells/mm³), a disseminated presentation is more likely.

Skin and oral disease
These are common and vary from being trivial to indicating life-threatening disseminated infection or malignancy. Most patients are affected at some time and type and severity often depend on the height of the CD4 count.

- Common skin problems are seborrhoeic dermatitis, xeroderma, itchy folliculitis, scabies, tinea, herpes zoster and papillomavirus infections
- Frequent oral or mucocutaneous lesions include oral and vaginal candidiasis, OHL, aphthous ulcers, herpes simplex and gingivitis
- In more advanced HIV, Kaposi's sarcoma

All patients	CD4 <200 cells/mm³
HBV surface antigen*	Chest X-ray
HBV core antibody†	HCV RNA
HCV antibody	Cryptococcal antigen
HAV IgG antibody	Stool OCP
Toxoplasma antibody	
Cytomegalovirus IgG antibody	CD4 <100 cells/mm³
Treponema serology	Cytomegalovirus PCR
Chest X-ray‡	Dilated fundoscopy
GUM screen	ECG
Cervical cytology (women)	Mycobacterial blood cultures

HAV, hepatitis A; HBV, hepatitis B; HCV, hepatitis C.
* HBV e antigen/antibody and HBV DNA if positive.
† HBV surface antibody if negative and history of immunization.
‡ If contact/past history of tuberculosis, injection drug users and those from tuberculosis-endemic areas.

Table 13.3 Baseline investigations.

(cutaneous and oral), molluscum contagiosum, severe and chronic mucocutaneous herpes simplex and CMV ulcers (oral) not uncommonly occur. Bacillary angiomatosis is unique to HIV and is due to *Bartonella* infection.

Gastrointestinal disease (Table 13.4)

HIV-related disease often involves the gastrointestinal (GI) tract. Loss of weight and appetite are common symptoms whatever the pathology.

• *Oesophageal disease* usually presents with pain on swallowing and dysphagia. Candidiasis is the cause in 80% (occurring in 30% of patients with

Table 13.4 Causes of gastrointestinal disease.

OCP). Pseudomembranous plaques are visible on barium swallow as filling defects (Fig. 13.1) and at endoscopy

• *Small bowel disease* is often associated with large-volume watery diarrhoea, abdominal pain and malabsorption. Where immunodeficiency is moderate (100–200 CD4 cells/mm³), *Cryptosporidium*, microsporidium and *Giardia* are the likely causes. When the CD4 level is <50 cells/mm³, *Mycobacterium avium-intracellulare* (MAI) and CMV are alternative diagnoses

• *Large bowel disease* presents with small-volume (often bloody) diarrhoea associated with abdominal pain. Any of the standard bacterial enteric pathogens may be responsible, as well as *Clostridium difficile*. CMV colitis is an important diagnosis in patients with low CD4 counts, occurring in up to 5% of patients. Diagnosis is by endoscopy, which often shows seg-

Site	Main causes of disease
Oesophagus	Candida, herpes simplex, CMV, aphthous ulcers, KS
Small bowel	MAI, CMV, Cryptosporidium, Microsporidium, Giardia
Large bowel	Salmonella, Campylobacter, Shigella, CMV, Clostridium difficile, Cryptosporidium
Biliary tract	Cryptosporidium, CMV, Microsporidium
Liver	Hepatitis B, hepatitis C, CMV, HIV drugs

CMV, cytomegalovirus; KS, Kaposi's sarcoma; MAI, disseminated *Mycobacterium avium-intracellulare*.

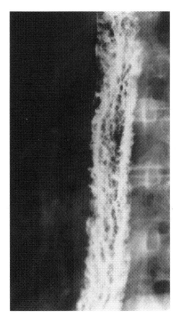

Fig. 13.1 Barium swallow of an AIDS patient with dysphagia showing multiple filling defects.

mental or confluent shallow or deep ulcers, and biopsy. Toxic megacolon, haemorrhage and perforation may complicate infection.

Hepatobiliary disease

• *Biliary disease* may complicate CMV, *Cryptosporidium* or microsporidium infection in the form of sclerosing cholangitis or acalculous cholecystitis. Presentation is with right upper quadrant pain, vomiting and fever; jaundice is infrequent. In sclerosing cholangitis, a rise in

Table 13.5 Differential diagnosis of CXR appearances.

serum alkaline phosphatase and γ-glutamyl transferase usually precedes the development of jaundice. Ultrasound imaging demonstrates biliary tract dilatation. However, endoscopic retrograde cholangiopancreatography (ERCP) is necessary to demonstrate the characteristic intrahepatic and extrahepatic beading of sclerosing cholangitis

• *Liver disease* may result from coinfection with HBV or HCV (see Chapter 10), or antiretroviral (ARV) drugs. Hepatitis B or C coinfections are becoming increasing problems in HIV. For both, viraemia is higher and disease more aggressive. In HBV coinfection, the immunosuppression seen in more advanced disease gives some protection, as hepatic damage is immune mediated. Both immune stimulants (interferon) and antivirals (3TC, tenofovir) have a place in treatment. In hepatitis C, the response to interferon and ribavirin is not as good as in HIV-negative persons.

Pulmonary disease

More than half of patients with HIV will develop pulmonary disease at some time. Several factors influence the likely cause including CD4 count, ethnicity and age, risk group, and any history of PCP prophylaxis.

Differential diagnosis

See Table 13.5.

• *Pneumocystis carinii* pneumonia (PCP) is discussed later in this chapter, and tuberculosis in Chapter 14

• *Bacterial pneumonia* due to *Streptococcus pneumoniae* is 150-fold commoner in HIV patients. It can occur at any CD4 count, is more commonly bacteraemic, and is a particular problem in

CXR appearance	Main causes
Diffuse infiltrate	*Pneumocystis carinii* pneumonia, tuberculosis (miliary), KS, NHL
Nodules/focal consolidation	KS, tuberculosis, NHL, pyogenic bacterial pneumonia
Hilar lymphadenopathy	Tuberculosis, KS, NHL
Pleural effusion	KS, tuberculosis, pyogenic bacterial pneumonia

KS, Kaposi's sarcoma; NHL, non-Hodgkin's lymphoma.

sub-Saharan Africa and injection drug users. All patients should be immunized with 23-valent polysaccharide vaccine although protection is proportional to CD4 count. Chest X-ray appearances may be atypical. Infections usually respond to standard antibiotic therapies. *Pseudomonas* and *Nocardia* infections are more likely in later-stage disease.

Nervous system/eye disease

Disease of the nervous system is common in HIV infection. Broad categories of presentation are space-occupying lesion, a global dementia illness, and root and peripheral nerve disease.

Differential diagnosis

See Table 13.6.

• Cerebral toxoplasmosis, progressive multifocal leucoencephalopathy, cryptococcal meningitis and CMV retinitis are discussed later in this chapter

• *AIDS dementia* incidence has dropped significantly since the introduction of HAART. It is characterized by deterioration in cognition, changes in affect, paraparesis, ataxia and incontinence: depression or psychosis may be the major feature. It is commoner in older patients with high plasma and cerebrospinal fluid (CSF) VL, and lower CD4 counts. Brain scanning re-

Table 13.6 Differential diagnosis of nervous system and eye disease.

veals diffuse cerebral atrophy. HAART may halt or even reverse disease progression

• *Vacuolar myelopathy* is a slowly progressive myelitis that can result in paraparesis

• *Peripheral neuropathy* from axonal degeneration can result from HIV or from the NRTI class of drugs (especially ddC, d4T and ddI). It is a predominantly distal sensory neuropathy of the lower limbs affecting up to 30% of patients and characterized by hyperaesthesia and pain on the soles of the feet, diminished light touch and vibration sensation, and loss of ankle reflexes

• *Polyradiculitis* due to CMV with rapidly progressive flaccid paraparesis, saddle anaesthesia, loss of reflexes and sphincter function may occur in late-stage HIV infection

• *Proximal myopathy* can rarely result from HIV or ZDV.

Disseminated and miscellaneous conditions

In the face of profound immune depletion (CD4 <50 cells/mm^3), disseminated disease is not infrequent and multiple OI pathogens may be identified (e.g. MAI, CMV). Often, presentation is with non-specific symptoms of fever and weight loss with evidence of anaemia on laboratory testing.

• MAI is discussed later in this chapter

• *Bone marrow.* Anaemia is not uncommon in late-stage HIV. Causes are many but marrow infiltration (e.g. MAI, NHL), marrow suppression

Site of disease	Presentation	Main causes
Brain	Space-occupying lesion	*Toxoplasma*, PCNSL, PMFL
	Encephalopathy	HIV, CMV
Meninges	Headache, meningitis	HIV seroconversion, *Cryptococcus*
Spinal cord	Spastic paraparesis	HIV vacuolar myelopathy
Nerve root	Weakness/numbness in legs, incontinence	CMV
Peripheral nerves	Pain, numbness in legs	HIV, drugs (ddC, d4T, ddI)
Retinitis	Floaters, field defects	CMV, toxoplasmosis, retinal necrosis (herpes simplex, VZV)
	Asymptomatic	HIV

CMV, cytomegalovirus; PMFL, progressive multifocal leucoencephalopathy; PCNSL, primary CNS lymphoma; VZV, varicella zoster virus.

(ZDV), blood loss (gastrointestinal Kaposi's sarcoma) and malabsorption (*Cryptosporidium*) represent the commonest. Leucopenia is usually seen in the context of marrow replacement as above or drug toxicity. Lymphopenia is a marker for HIV and immunological function. Thrombocytopenia can appear early (5–10%) with a presentation similar to ITP: response to immunoglobulin is good but short-lived; the treatment of choice is HAART

• *HIV-associated nephropathy* (HIVAN) is the most important renal condition, being seen most commonly in Africans, men and injection drug users, and presenting with nephrotic syndrome, chronic renal disease or a combination of both. HAART and high-dose steroids may have a place in the acute management; however, many patients end up requiring dialysis. Also, several drugs used in HIV management can cause renal disease (e.g. indinavir, tenofovir)

• *Heart disease* resulting from HIV which is symptomatic is rare (<3%). More important is the probable association of an increased risk of ischaemic heart disease in patients on HAART, particularly the protease inhibitors

• *Hypoadrenalism* occurs in up to 25% of patients with late-stage HIV infection.

Specific diseases

Cryptosporidium and microsporidium

• *Cryptosporidium* accounts for 15–20% of cases of diarrhoea; at lower CD4 counts, the infection becomes persistent with malabsorption and marked weight loss. Diagnosis is confirmed on stool microscopy in 90% with acid-fast or immunofluorescence stains. Patients with CD4 counts below 200 cells/mm³ should be advised to boil drinking water and minimize contact with animals

• Infection with microsporidium presents identically to that with *Cryptosporidium* although it is not usually as severe. Three main species infect man: *Enterocytozoon bieneusi*, *Encephalocytozoon hellem* and *Encephalocytozoon intestinalis*. E.

intestinalis can be treated with albendazole but reconstitution of the immune system with HAART offers the only chance of cure for patients infected with *E. bieneusi*. Diagnosis is on stool microscopy and duodenal biopsy. Occasionally, electron microscopy is necessary to confirm the presence of infection and is the only way to speciate microsporidium.

Pneumocystis carinii pneumonia (PCP)
Epidemiology
This is the commonest AIDS-defining diagnosis in the Western world, especially in newly presenting patients unaware of their positive status. Like other opportunistic infections in HIV, its incidence has significantly decreased (72% in 5 years) because of HAART and the use of primary prophylaxis.

Pneumocystis carinii
• Genotypically a fungus but phenotypically a protozoan: it cannot be cultured, but the cyst and trophozoite are easily stained
• Results in PCP by reactivation of latent infection in the majority of patients: reinfection may also occur
• Transmitted by the airborne route with the source being other infected humans; however, environmental reservoirs have not been excluded
• Causes asymptomatic infection that is highly prevalent among the general population and occurs at an early age
• Rarely causes disease outside of severe immunocompromise.

PCP
• Risk is inversely correlated with CD4 count: 40% of patients with a count of <100 cells/mm³ not taking prophylaxis will develop the disease annually
• Can be effectively prevented when daily co-trimoxazole (the drug of choice) is the prophylactic agent; dapsone, atovaquone or aerosolized pentamidine inhaled 2–4 weekly are effective alternatives. This should be started when the CD4 count is <200 cells/mm³.

Clinical features

Typical features are:
- 2–3-week history of fever, dry cough and disproportionate shortness of breath with 'wet' crackles in severe disease
- failure to respond to standard antibiotics
- other clinical markers of HIV (OHL, mucocutaneous herpes simplex or oropharyngeal candidiasis)
- exercise-induced O_2 desaturation and arterial hypoxaemia
- a perihilar ground-glass appearance on CXR. In early disease, the CXR may be normal (15–20%) and in late disease, an adult respiratory distress syndrome (ARDS) picture is common (Fig. 13.2)
- complicating pneumothorax, bacterial superinfection and respiratory failure in severe disease.

Diagnosis

- Most infections are diagnosed clinically on the basis of characteristic symptoms, CXR and arterial hypoxia
- Specific diagnostic tests are cytology of nebulized hypertonic saline-induced sputum

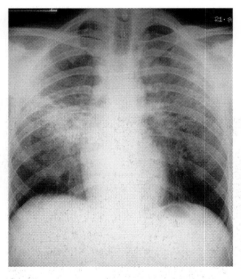

Fig. 13.2 Chest X-ray of a patient with *Pneumocystis carinii* pneumonia showing bilateral hazy midzonal opacities.

(30–80% sensitivity), bronchoscopy with lavage (90%) or together with transbronchial biopsy (95%).

Treatment

- Moderate to severe PCP (judged by hypoxaemia): high-dose parenteral co-trimoxazole with corticosteroids. Pentamidine or intravenous clindamycin and oral primaquine are alternatives
- Mild disease: oral high-dose co-trimoxazole, dapsone and trimethoprim, atovaquone, or daily inhaled pentamidine.

Prognosis

There is a 10–15% acute mortality. Poor prognostic factors include delayed diagnosis, not being on prophylaxis, low CD4 count, extensive CXR changes, low hypoxaemia ratio, high lactate dehydrogenase (LDH), hypoalbuminaemia and additional pulmonary infection.

Mycobacterium tuberculosis

See Chapter 14.

Cerebral toxoplasmosis

Epidemiology

Toxoplasma gondii (see p. 224 for more details):
- causes a mild glandular fever-like or subclinical illness in most infected immunocompetent persons
- results in latent tissue cysts being formed that persist for life
- has a prevalence varying between regions and increasing with age (e.g. UK seroprevalence of 8% in children <10 years rising to 47% in those >60 years)
- has a seroconversion rate of 0.5–1%/year
- will reactivate in the brain in 30% of *Toxoplasma*-seropositive patients with a CD4 count <100 cells/mm³ in the absence of primary prophylaxis
- localizes to the basal ganglia, cortex and brainstem, typically at the grey/white matter interface.

Clinical features and diagnosis

- One to two-week history of headache, fevers

and drowsiness followed by confusion and focal signs: one-third develop seizures.

- Supporting evidence:
  - positive *Toxoplasma* IgG serology (90%)
  - imaging with contrast-enhanced computed tomography (CT) or magnetic resonance (MR) scan showing multiple ring-enhancing lesions with marked surrounding oedema and, often, mass effect (Fig. 13.3)
- Confirmation:
  - empirical trial of anti-*Toxoplasma* therapy with clinical and scan improvement occurring within the first week
  - positive CSF PCR for *Toxoplasma* (although a lumbar puncture is usually contraindicated because of raised intracranial pressure, ICP)
  - brain biopsy.

Treatment

- Pyrimethamine (and folinic acid) with sulpha-diazine (or clindamycin) for 8 weeks
- Dexamethasone if there is significant mass effect.

Prevention and prognosis

- HIV patients should avoid undercooked meats and contact with cat faeces without protection
- Where the CD4 count is <100 cells/mm³, primary prophylaxis should be given with co-trimoxazole or dapsone and pyrimethamine in seropositive patients
- Significant neurological morbidity occurs in 15% with occasional mortality.

Primary CNS lymphoma

See p. 182.

Progressive multifocal leucoencephalopathy (PMFL)

Epidemiology and clinical features

- A fatal demyelinating disease caused by JC papovavirus
- Occurs in 2–3% of AIDS patients with CD4 counts <50 cells/mm³
- Due to reactivation of childhood infection (90% of adults seropositive)
- Results in a rapidly progressive demyelinating disease
- Usually presents with visual field defects, ataxia or hemiparesis.

Diagnosis and treatment

- Best visualized by MRI which shows high-intensity white matter signals on T2-weighted images. Contrast enhancement is rare and mass effect not seen
- Definitively diagnosed by brain biopsy. However, the MRI scan and clinical picture are so characteristic that this is rarely necessary
- Has a very bad prognosis with the majority of patients progressing to death over a few months
- Has occasionally responded to HAART with or without cidofovir.

Cryptococcal meningitis

Epidemiology and clinical features

Cryptococcus neoformans

- Is a budding encapsulated yeast

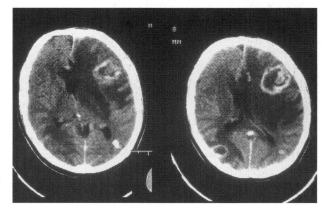

Fig. 13.3 CT scans of brain of an HIV-positive patient with cerebral toxoplasmosis showing two hypodense shadows with contrast-enhanced rings (picture on right) and significant improvement after 2 weeks of treatment (left).

• Is the commonest cause of meningitis in HIV when the CD4 count is usually <100 cells/mm³)
• Rarely causes non-meningeal or disseminated disease

Cryptococcal meningitis
• Presents with a 2–3-week history of headache, fever, vomiting and mild confusion; neck stiffness is often absent (<25%), and seizures, papilloedema and focal features rare; 10% of patients are asymptomatic
• May be complicated by severe raised ICP, deafness and blindness.

Diagnosis and treatment
• Characteristically results in a raised-protein, lymphocytic CSF with a low glucose. However, the CSF findings may be normal
• Associated with numerous organisms in the CSF diagnosed by India ink stain (70%), by culture (>95%), and by high-titre CSF and/or serum cryptococcal antigen (95%)
• May also be cultured from blood, urine, gut or bone marrow
• 10–20% of patients require treatment for raised ICP with repeated lumbar punctures, acetazolamide or shunting
• Associated with significant mortality (5%) and morbidity
• Treated with amphotericin B and 5-flucytosine for 2 weeks followed by 6–8 weeks high-dose fluconazole.

CMV retinitis
Epidemiology of CMV infection
• A member of the herpesvirus group, CMV is found worldwide. Between 60 and 90% of adults carry antibodies in their blood, signifying previous infection through contact with bodily secretions from an infected carrier (commonly intimate contacts)
• Infection rate is higher in poorer socioeconomic groups and in sexually active people
• After primary infection (usually asymptomatic), the virus persists in the body in a latent form
• Occasionally, reactivation occurs and the virus

appears in saliva, urine, cervical secretions, semen and breast milk, thereby facilitating transmission to others. Congenital infection may occur *in utero* or during birth (see Table 16.1, p. 204)
• In the severely immunocompromised, clinical disease is common, usually from reactivation, but primary infection may occur in a seronegative recipient of blood or organ from a seropositive donor and may present as retinitis, pneumonitis (usually in bone marrow or transplant recipients), gut mucosal ulcerations, hepatitis, cholangitis, encephalitis or neuropathies
• CMV retinitis:
  • is now uncommon in HIV as a result of HAART (90% reduction in 5 years)
  • presents with floaters, flashing lights, field defects or rarely diminished central visual acuity
  • diagnosis is clinical by fundoscopy: usually reveals haemorrhagic exudates along retinal vessels or a well-demarcated peripheral V-shaped area of disease
  • is bilateral in 30% at presentation
  • is not associated with any recovery of vision despite successful treatment
  • may be complicated by retinal detachment because of necrosis.

Treatment
• Must be treated aggressively with parenteral ganciclovir (foscarnet, cidofovir or oral valganciclovir are alternatives), followed by a reduced maintenance dose. An intravitreal ganciclovir implant may be an alternative for some
• May be complicated by an immune vitritis on HAART
• Usually remains inactive despite discontinuation of maintenance treatment where there has been a CD4 rise above 200 cells/mm³
• Must be monitored after discontinuation of maintenance.

Disseminated *Mycobacterium avium-intracellulare* (MAI) infection
Epidemiology and clinical features
Mycobacterium avium-intracellulare is an environ-

mental mycobacterium commonly found in water and food. Infection follows on from colonization of the respiratory and gastrointestinal tracts in most instances. Disseminated disease:
• only occurs when the CD4 count is <50 cells/mm³
• has fallen in incidence by over 90% since the advent of HAART
• affects all organs (particularly the reticuloendothelial system) with massive infiltration of organisms and minimal inflammatory response
• presents with fever, sweats, weight loss, chronic diarrhoea, vomiting and abdominal pain; hepatosplenomegaly and lymphadenopathy are usual on examination. CT scanning usually shows intra-abdominal and mediastinal lymphadenopathy.

Diagnosis and treatment
• Often causes anaemia and a raised alkaline phosphatase
• Is confirmed as mycobacterial by finding numerous acid-fast bacilli (AFB) in induced sputum, bone marrow, liver or duodenal biopsy. MAI is confirmed on positive culture from these sites or from blood (positive in 91% of patients and the usual source) or stool, and speciation using DNA probes. Rarely, other non-tuberculous mycobacteria have been described in HIV with similar presentation (*M. haemophilum* and *M. genavensae*)
• Treated with a minimum of four drugs: usually ethambutol, azithromycin (or clarithromycin), rifabutin and ciprofloxacin. With response to HAART (CD4 count >100 cells/mm³), multidrug therapy can often be stopped and replaced with azithromycin prophylaxis
• May be a cause of immune reconstitution syndrome (IRIS) in patients who respond to HAART where focal disease (e.g. joint, lymph node) has been described
• Is less likely when weekly azithromycin in those with CD4 counts <50 cells/mm³ is used as prophylaxis.

HIV-related tumours
Kaposi's sarcoma (KS) (1000-fold higher rate)

and NHL (60-fold higher rate) are the most important complicating HIV-related malignancies. An increased incidence of Hodgkin's lymphoma (fivefold) and anogenital cancers (fivefold) is also seen. There are less strong associations with other cancers (e.g. conjunctival squamous cell carcinoma and seminoma). Although cervical *in situ* cancer (CIN III) is more common in HIV, there has been no increased incidence of cervical cancer.

Kaposi's sarcoma
Epidemiology
• A rare tumour in HIV-negative persons, although long recognized in elderly males of Mediterranean or Jewish descent, transplant recipients, and children and young adults in sub-Saharan Africa
• Much commoner in men who have sex with men, bisexuals and those from sub-Saharan Africa than in other risk groups
• Mainly transmitted sexually but probably also through saliva: it can also be transmitted vertically
• Associated with human herpesvirus 8 (HHV-8). Tissue virus and HHV-8 antibodies can be identified in nearly all patients with KS. Also, seroprevalence studies show higher prevalence of HHV-8 antibodies in those groups more at risk of the disease (20–50% men who have sex with men, individuals from Africa and populations in Mediterranean Europe). In coinfected male homosexuals, the 10-year cumulative risk for developing KS is 30–50% with a median time to development of 3–5 years. HHV-8 contains several candidate growth factor and oncogenes
• Consists of spindle cells, endothelial cells, fibroblasts and inflammatory cells.

Clinical features
• Lesions vary tremendously in appearance but are usually discrete, raised, purple, non-tender papules or nodules that are often symmetrically arranged along crease lines. Common sites are trunk and lower limbs
• Oedema from lymphatic involvement can occur

- With disease progression, the lesions become more numerous and larger
- Progression is often indolent but may be fulminant with rapid visceral involvement. This is more likely with lower CD4 counts
- Oral KS typically involves the palate, gum margins and fauces. It is a strong predictor of GI and pulmonary disease
- Hepatosplenomegaly usually indicates disease of these organs
- Pulmonary KS occurs in 10–15% of patients. It presents with breathlessness, cough, haemoptysis, chest pain and fever: it may mimic the CXR appearance of PCP. A pleural effusion is found in one-quarter.

Diagnosis and treatment
- Diagnosis is clinical in most cases. Histology is definitive
- For limited localized disease, surgical excision, intralesional chemotherapy, liquid nitrogen or laser therapy, radiation or retinoic acid gel can all be used. Also, on initiation of HAART, this form of KS often resolves spontaneously over 6 months
- For more widespread cutaneous or visceral KS, chemotherapy should be used
  - First-line treatment is cyclical liposomal doxorubicin (response rate 60%)
  - For refractory or relapsed disease, paclitaxel is the best drug
- With HAART, there has been a 68% fall in the incidence of KS. It is also important in maintaining remission. It is therefore a vital component of management
- Prognosis depends upon CD4 count and extent of disease.

HIV-associated non-Hodgkin's lymphoma
Epidemiology
- Usually occurs in late HIV and is the initial AIDS-defining illness in 2–3% of patients. Lifetime risk of developing NHL is 5–10%
- Seen in all HIV risk groups
- Has reduced in incidence by 40% since the introduction of HAART
- Is of B-cell lineage (95%) compromising several histological types: diffuse large cell (30%); small cleaved cell, Burkitt's and non-Burkitt's (30%); and large cell immunoblastic (25%)
- Is intimately associated with EBV, which is present in approximately 50% of non-Hodgkin's and nearly all Hodgkin's lymphomas. HHV-8 is found in body cavity (pleura, pericardium or peritoneum) lymphoma which accounts for <2% of NHL and the premalignant condition, Castleman's disease.

Clinical features
Weight loss, fever and night sweats are common, reflecting disseminated disease. Lymphadenopathy occurs in 75% and extranodal sites are involved in up to 60% (bone marrow, GI tract, CNS, liver, oral cavity, skin and lungs).

Diagnosis and treatment
- Diagnosis is invariably from biopsy of one of these sites
- On diagnosis, disease extent must be staged, requiring CT of the thorax, abdomen and pelvis, bone marrow aspirate and biopsy, and lumbar puncture for cytology; 20% of patients with leptomeningeal CNS NHL are asymptomatic
- Poor prognostic factors are CD4 <100 cells/mm^3, previous AIDS diagnosis, disseminated and/or neurological disease, raised LDH, age >35 years, not being on HAART, and intravenous drug use risk group
- Multiagent chemotherapy is used (e.g. CHOP or VAPEC-B), although lower doses are often employed. Prolonged remission over 24 months is seen in 20–30% of those patients with less than two poor-prognosis risk factors
- HAART improves chemotherapy response, event-free remission and overall survival.

Primary CNS lymphoma (PCNSL)
Epidemiology and clinical features
- Complicates late-stage HIV (CD4 <50 cells/mm^3)
- Occurs in 5% of AIDS patients and accounts for 20% of all focal CNS lesions
- Is a high-grade, diffuse, B-cell tumour (EBV is found in over 90%)

• Progresses fast (weeks to months) with features of raised ICP developing early and focal signs later

• Is visible as a single, large homogeneously enhancing lesion with surrounding oedema and mass effect on CT or MRI (greater sensitivity) scanning. Multiple lesions may be present (50%) but rarely number more than two or three.

Diagnosis and treatment
• Often causes CSF abnormalities but cytology is only positive in 25%

• Positive EBV DNA in CSF has high sensitivity and specificity for PCNSL

• Positron emission tomography (PET) scanning may distinguish lymphoma (which results in a 'hot spot') from other mass lesions

• Brain biopsy is the definitive investigation but is non-diagnostic in up to one-third. However, a trial of anti-*Toxoplasma* therapy should always be given; failure to improve clinically or on scanning is consistent with PCNSL

• Treatment is usually palliative with dexamethasone and symptomatic relief. Whole-brain irradiation may provide temporary respite

• Distinction from cerebral toxoplasmosis may be impossible, although in the absence of *Toxoplasma* antibodies, this diagnosis is much less likely. Life expectancy is measured in months.

Management of HIV

The aim of therapy is to maintain immunological health (CD4 count > 200 cells/mm^3), reduce the VL to an undetectable level, improve quantity and quality of life without unacceptable side-effects, and reduce transmission (e.g. from mother to child). Starting therapy is dependent upon the symptom status of the patient and their CD4 count. The VL provides information on speed of progression prior to treatment but its main value is to assess the effectiveness of a patient's drug regimen.

Drugs and treatment
• There are three major classes of drugs: nucleoside reverse transcriptase inhibitors (NRTIs), non-nucleoside reverse transcriptase inhibitors

(NNRTIs) and protease inhibitors (PIs). NRTIs and NNRTIs cause chain termination of the forming DNA strand by competing with natural nucleosides (e.g. zidovudine for thymidine), or binding near to the active enzyme site, respectively, and PIs interfere with the cleavage of polypeptides into the functional proteins. Very recently, a new class of drug, fusion inhibitors (which block viral entry into cells), has become available

• Highly active antiretroviral therapy (HAART) is a combination of two NRTIs with either a PI or an NNRTI. Occasionally other combinations are used. Higher trough levels of PI drugs are achieved with low-dose ritonavir boosting, which is now recommended when using these agents

• Several drugs are now licensed and there are numerous more in development (Table 13.7). Combination tablets are also available, reducing the tablet load (e.g. Combivir which is ZDV and 3TC)

• The importance of adherence to the drug regimen in achieving success has now been recognized. Once-daily regimens offer an easier schedule for many patients and several drugs can now be administered this way (e.g. tenofovir, ddI, 3TC, FTC, efavirenz and nevirapine). Ritonavir, because of its potency as an inhibitor of the P450 complex, allows boosting of other PIs and a daily dosing (e.g. saquinavir, lopinavir and fos-amprenavir).

The drug-naïve patient
• Treatment is recommended when the patient is symptomatic or the CD4 count has fallen below 250–300 cells/mm^3. Above this level, significant HIV-related events are rare and there is no evidence that early therapy is beneficial. Within this range, patients with high viral load, rapidly falling CD4 counts, or hepatitis C coinfection should be considered for earlier treatment. A typical combination would be either efavirenz or lopinavir/ritonavir with ZDV and 3TC. The nucleosides d4T and ddC and the combination of d4T and ddI are no longer recommended because of long-term toxicity

• The regimen must be individualized for the patient. Factors that must be considered are

NRTIs	NNRTIs	PIs	Others
ddC	Nevirapine	Indinavir	T-20 (fusion inhibitor)
ddI	Efavirenz	Saquinavir (soft gel)	Tenofovir
3TC	Delavirdine	Ritonavir	Hydroxyurea
ZDV		Nelfinavir	
D4T		Lopinavir/ritonavir	
Abacavir		Fos-amprenavir*	
FTC*		Tipranavir*	
		Atazanavir*	

NRTI, nucleoside reverse transcriptase inhibitor; NNRTI, non-nucleoside reverse transcriptase inhibitor; PI, protease inhibitor.
* Unlicensed.

Table 13.7 Antiretroviral drugs.

the fit of the drug regimen to the patient's lifestyle, the stage of disease, any coexisting or past medical history, the possibility of additive side-effects (e.g. ddI or d4T and neuropathy), the potential for drug interactions with non-HIV medications (e.g. rifampicin), antagonistic NRTI combinations (e.g. ZDV and D4T should not be given together), and CNS penetration
• CD4 and viral load should be monitored on a regular basis. Failure to achieve an undetectable viral load within 12–24 weeks could indicate poor adherence, low drug absorption, drug interaction, suboptimal combination or primary resistance of the virus
• A viral resistance test should be performed when the patient is pregnant, there has been recent seroconversion, the partner is on antiretrovirals or the viral load fails to fall as expected on HAART. Primary resistance is uncommon in Western countries (5–15%).

The drug-experienced patient
• Switching a patient's therapy may be necessary because of virological failure, drug toxicity but with good virological control, or to simplify a regimen
• Where there has been virological failure, current and past HIV resistance tests are important guides as to what future combination is likely to be most active. Similarly, knowledge of past combinations and viral control on these is crucial. Wherever there are drugs that the patient has not received, it is important to try and preserve these to form part of a new combination with as many new or active drugs as possible rather than adding sequentially to a failing regimen.

Drug side-effects
These are not infrequent and may be class specific (usually occurring late) or drug specific (usually early).
• For NRTIs, mitochondrial dysfunction is thought to play a major role in adverse reactions. Important side-effects are peripheral neuropathy (ddC, d4T and ddI), hepatic steatosis and lactic acidosis (d4T, ddI), pancreatitis (ddI), anaemia (ZDV) and peripheral fat atrophy (d4T). Hypersensitivity to abacavir can also occur in 3%. Tenofovir, which is a nucleotide RTI, may rarely cause renal tubular dysfunction, especially when given with other nephrotoxic drugs
• Rash is the major class-specific side-effect for NNRTIs, with hepatitis (nevirapine) and CNS side-effects (efavirenz) being drug specific
• The major class side-effects for PIs are fat redistribution (mainly central fat accumulation) occurring in up to 20% at 2 years, insulin

resistance with abnormal glucose tolerance, and hyperlipidaemia. Important drug-specific side-effects include renal stones with indinavir and diarrhoea with nelfinavir

• Lipodystrophy is characterized by peripheral fat wasting (cheeks, limbs, buttocks), localized collections (buffalo hump) and central adiposity and results from both PIs and NRTIs.

Children

• The principles for drug management of HIV-positive children are the same as for adults

• Children should be treated if they are symptomatic or have a low CD4 count (or CD4 percentage when ≤6 years). The height of the VL is indicative of the speed of progression so is a good predictor as to when therapy is going to be indicated

• Achieving an undetectable VL is more of a challenge in children because VLs tend to be higher, adherence is more difficult, and not all antiretroviral drugs are available in a suitable formulation

• Family-centred systems of care are necessary to support the child and parents

• As in adults, PI-based or NNRTI-based HAART should be used; any of the NRTIs can be used but with the same restrictions applying to adults when treating children naïve to drugs

• Virological failure occurs in 50% at 1 year. Treatment changes are decided on the same basis as they are for adults, including obtaining a resistance assay.

Perinatal transmission

• All patients should be treated in an effort to prevent transmission

• Without treatment, the risk of MTCT is 14%; this falls to <1% when the maternal VL is <1000 copies/mL

• Patients not on therapy who become pregnant should not be commenced on treatment until 12 weeks when the major organ development has taken place. With a few exceptions, those on treatment when they discover they are pregnant should continue with the same combination

• A combination of ZDV, 3TC and nevirapine is the most commonly prescribed combination, with a good safety record in pregnancy

• For perinatal prophylaxis, nevirapine should be given as a single dose to the mother at the onset of labour and to the newborn within 48 h. ZDV should be commenced as an IV infusion at the onset of labour and the neonate should be treated for 4–6 weeks

• All patients should be offered a Caesarean section, although there is no evidence it provides additional benefit when the viral load is undetectable at term

• In mothers with a low viral load who do not warrant treatment themselves, an alternative is commencing ZDV at 16 weeks together with delivery by Caesarean section.

Postexposure prophylaxis

• Postexposure prophylaxis (PEP) can be given after occupational or sexual exposure (e.g. condom breakage in HIV-serodiscordant partners, victims of rape)

• The risk of HIV from a needle-stick exposure is 0.39%. It is increased where there is a deep injury, visible blood on the device, when it has been sited in an artery or vein, and when the source patient has AIDS or a high viral load

• Combination therapy is recommended after careful assessment of the risk: the first dose should be given as soon as possible

• Protection is not absolute and reports of seroconversion despite taking a full course of three drugs started within hours of the exposure have been described

• Approximately 77% of persons receiving PEP experience side-effects and only 40% complete therapy

• Recommended PEP is ZDV, 3TC and indinavir or nelfinavir for 28 days. Review of the PEP combination is necessary in the light of the source patient's treatment history, viral load and any sensitivity test results.

Prevention of infection

• All patients should be screened for immunity to hepatitis A and B and vaccinated if unprotected. HBV surface antibodies levels need to be monitored and boosters given when <100 IU/mL

• Pneumococcal vaccine (every 5 years) and influenza vaccine (every year) should be given to all patients

• Response to all immunizations is significantly less when the CD4 count is <200 cells/mm^3

• Live attenuated vaccines should be avoided completely (BCG), or given only to those with relatively preserved or restored immune function (e.g. oral polio, yellow fever). MMR vaccine is safe and should be given

• Primary prophylaxis is started at certain CD4 count levels when there is a risk of infection occurring: 200 cells/mm^3 for PCP, 100 cells/mm^3 for *Toxoplasma* and *Cryptococcus*, and 50 cells/mm^3 for MAI. Secondary prophylaxis is commenced post successful treatment of the OI, usually with the same drugs used to treat the infection but at lower doses

• Prophylaxis can be stopped when the CD4 threshold at which they were introduced is reached

• Vaccine development is slow. The massive viral turnover and the frequent generation of antigenically distinct variants provide a significant challenge

• Useful websites:
  • www.aidsmap.com
  • www.i-Base.info.

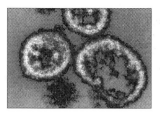

Tuberculosis

Epidemiology

Tuberculosis (TB) is the most common infectious disease in the world, with an estimated one-third of the population infected and 2.5 million deaths annually. A falling incidence has been reversed, with increases in both developed and developing nations since the mid-1980s: human immunodeficiency virus (HIV) fuels much of the new epidemic. *Mycobacterium tuberculosis*:

• infected 8.7 million new cases in 2000 with a global incidence rate that increases by 0.4% per year. The bulk of new infections are in southeast Asia (3 million) and Africa (2 million). One-third of patients with tuberculosis in Africa are coinfected with HIV. By 2005, the World Health Organization predict there will be 10.2 million new cases and Africa will have more cases than any other region (up 10% annually). In the UK numbers are increasing, with London seeing a 40% increase between 1999 and 2000

• is commoner in certain conditions where susceptibility is increased: HIV, silicosis, immunocompromise, malignancy (especially leukaemia and lymphoma), insulin-dependent diabetes mellitus, chronic renal failure and gastrointestinal (GI) disease with malnutrition

• is increasing as a result of many factors—see Table 14.1

• is seen more frequently in children, close con-tacts of patients with smear-positive pulmonary tuberculosis, persons with chest X-ray (CXR) evidence of healed tuberculosis, and where primary infection occurred < 1 year previously

• causes pulmonary (75%) and extrapulmonary disease (lymph node, bone and joint, meningeal, pericardial, genitourinary and GI tract)

• can result in occasional large outbreaks in institutions (schools, paediatric wards)

• is transmitted by aerosol from smear-positive pulmonary cases: 10% of close contacts develop primary tuberculosis, 5% of those infected develop progressive primary infection, and another 5% will reactivate in later life (postprimary tuberculosis)

• is rarely infectious in primary disease; in immunocompetent adults, tuberculosis is usually a result of reactivation (70%)

• in HIV-positive persons results in a greater likelihood of contracting disease if exposed (30%), of developing progressive primary disease (30%), of developing extrapulmonary and atypical pulmonary manifestations (50%), of being due to reinfection (50%) and of reacting to standard drugs

• can be resistant to one or more of the antituberculosis drugs. Primary and secondary (past treatment) resistance is uncommon in the UK (< 5%) but is frequent in many developing countries. Multidrug-resistant strains (MDRTB) are usually seen in previously suboptimally treated patients from abroad and in HIV care facilities

Developed countries	Developing countries
HIV[1]	HIV
Immigration from high-prevalence areas[2]	Population increase[4]
Increasing life expectancy of the elderly	Lack of access to health-care
Social deprivation	Poverty, civil unrest
Drug resistance (MDRTB)[3]	Ineffective control programmes
	Drug resistance (MDRTB)[3]

1 Mainly urban; 3% of cases in 1998 were coinfected in the UK.
2 One-third of new cases in the UK in 1998 were born outside the country.
3 Rifampicin/isoniazid resistance (with/without additional drugs). MDRTB, multidrug-resistant tuberculosis.
4 75% predicted increase in India over 30 years.

Table 14.1 Factors resulting in an increase in tuberculosis.

• has an incubation period (IP) for primary infection of 4–16 weeks
• is one of a large number of mycobacterial species (mostly environmental), many of which can also cause disease in humans—see Table 14.2.

Pathology and pathogenesis

Following inhalation of M. tuberculosis, the disease progresses as follows.

Table 14.2 Non-tuberculous mycobacteria causing human disease.

• A small subpleural lesion, termed a Ghon focus, develops
• Infection then spreads to the hilum and mediastinal lymph glands to produce a primary complex. These enlarge with an inflammatory granulomatous reaction, which may caseate
• In 95% of cases, the primary complex heals spontaneously in 1–2 months, sometimes with calcification, and the individual becomes tuberculin skin test positive
• In 10–15%, the infection spreads from the primary complex, locally to a bronchus causing pressure on (collapse, obstructive emphysema) or rupture into (endobronchial, bronchopneumonia) a bronchus, via the lymphatics to the pleura causing a pleural effusion, or via the bloodstream to cause disseminated lesions

Pulmonary	Lymph node	Soft tissue/skin	Disseminated[1]
M. bovis/M. africanum[2]	MAC[3]	M. leprae	MAC[5]
M. xenopi	M. bovis[2]	M. ulcerans[4]	M. haemophilum
M. kansasii	M. malmoense	M. marinum	M. genavensae
M. malmoense	M. fortuitum	M. fortuitum	Bacille Calmette–Guérin (BCG)
MAC	M. chelonei	M. chelonei	

1 Seen in immunodeficiency states.
2 Cause 'classical' tuberculosis. M. bovis reservoir is in cattle. M. africanum is restricted to west/central Africa.
3 Mycobacterium avium complex (M. scrofulaceum, M. intracellulare and M. avium).
4 Causes Buruli ulcer, prevalent in Africa, northern Australia, and south-east Asia.
5 HIV-associated.

- In some, disease then progresses with the development of miliary or meningeal tuberculosis. In others, dormant foci are created in the bone, lungs, kidneys, etc., which reactivate in later life
- Occasionally, the tonsil, intestine and skin may be the site of primary disease
- The virulence factors of *M. tuberculosis* have not been fully elucidated. The organism is versatile, with the ability to multiply rapidly outside cells within cavities, to survive inside macrophages and prevent fusion of the lysosome and phagosome, and to survive in a relatively inactive state with only infrequent bursts of division.

Clinical features

Pulmonary disease

- The primary pulmonary complex is often asymptomatic or marked only by a self-limiting febrile illness. Clinical disease results either from the development of a hypersensitivity reaction or from the infection pursuing a progressive course. Erythema nodosum may be the presenting feature and is associated with a strongly positive tuberculin skin test. Progressive primary disease may appear during the course of the initial illness or after a latent interval of weeks or months. Endobronchial tuberculosis may result in wheezing and cough. Collapse and/or consolidation from obstruction or tuberculous bronchopneumonia is usu-

ally associated with constitutional symptoms, cough and sputum
- Miliary tuberculosis is a severe infection, often diagnosed late. The patient usually has a 2–3-week history of fever, night sweats, anorexia, weight loss and a dry cough. Hepatosplenomegaly may be present (25%). Auscultation of the chest may be normal but with advanced disease widespread crackles are evident. Choroidal tubercles occur in 5%. The CXR shows fine 1–2-mm 'millet-seed' appearances through both lungs
- Cryptic tuberculosis is often associated with a normal CXR and affects the elderly, presenting with unexplained weight loss, fevers, hepatosplenomegaly and blood film abnormalities and is usually confirmed by liver or bone marrow biopsy
- Postprimary adult pulmonary tuberculosis is the most frequent presentation. Cough, haemoptysis, dyspnoea, anorexia and weight loss associated with fevers and sweats is typical. The history is subacute (4–8 weeks) in the majority. Auscultation of the chest usually reveals localized signs
- In HIV where the CD4 count is >350 cells/mm^3, disease is more likely to be reactivated upper lobe open cavitatory disease; as immunosuppression increases, miliary, atypical pulmonary and extrapulmonary (especially pericardial, abdominal and meningeal) tuberculosis become progressively more common, as does mycobacteraemia. Constitutional symptoms of fever and night sweats are usually present. MDRTB in HIV centres caused major nosocomial outbreaks initially in the USA in the early 1990s with very high mortalities. These are now rare in developed countries.

Table 14.3 Complications of pulmonary tuberculosis.

Acute	Chronic
Respiratory failure, adult respiratory distress syndrome	Aspergilloma
Haemoptysis (occasionally massive)	Lung fibrosis, cor pulmonale
Pleural effusion, empyema	Lung/pleural calcification
Pericardial effusion	Amyloidosis
Laryngitis	Atypical mycobacterial colonization (e.g. *M. malmoense*)

Complications

See Table 14.3.

Lymphadenitis

• The commonest site of extrapulmonary disease
• Disease may represent primary infection, spread from contiguous sites or reactivation of infection
• Cervical and mediastinal glands are affected most commonly, followed by axillary and inguinal: in 5% more than one regional group is involved
• Constitutional disturbance and evidence of associated tuberculosis are usually lacking
• In non-immigrant children in the UK, most mycobacterial cervical lymphadenitis is caused by environmental (atypical) mycobacteria, especially of the M. avium complex
• The nodes are usually painless and initially mobile but become matted together with time. When caseation and liquefaction occurs, the swelling becomes fluctuant and may discharge through the skin with the formation of a collar-stud abscess and sinus formation
• During or after treatment, paradoxical enlargement, development of new nodes or suppuration may all occur but without evidence of continued infection; rarely surgical excision is necessary.

Gastrointestinal disease

• Tuberculosis can affect any part of the bowel and patients may present with a wide range of symptoms and signs
• Ileocaecal disease accounts for half the cases of abdominal tuberculosis. Fever, night sweats, anorexia and weight loss are usually prominent and a right iliac fossa mass may be palpable; up to 30% present with an acute abdomen. Abdominal scanning may reveal thickened bowel wall, lymphadenopathy, mesenteric thickening or ascites. Barium and small bowel enemas may reveal narrowing, shortening and distortion of the bowel with caecal involvement predominating. Differentiation is from Crohn's disease
• Tuberculous peritonitis is associated with

fevers, abdominal pain and distension. Exudative ascites is common, or there may be palpable masses of matted omentum and loops of bowel
• Diagnosis of abdominal tuberculosis rests on obtaining histology by either colonoscopy (ileocaecal disease), laparoscopy (peritoneal) or minilaparotomy
• Low-grade hepatic dysfunction is common in miliary and cryptic tuberculosis, which often presents as a pyrexia of unknown origin, when biopsy reveals granulomata. Occasionally patients may present frankly icteric with a mixed hepatic/cholestatic picture.

Pericardial disease

• Disease occurs in two main forms: pericardial effusion and constrictive pericarditis
• Fever and night sweats are rare and presentation is insidious with breathlessness and abdominal swelling
• Pulsus paradoxus, a very raised jugular venous pressure, hepatosplenomegaly, prominent ascites and the absence of peripheral oedema are common to both
• Pericardial effusion is associated with increased pericardial dullness and a globular enlarged heart on CXR whereas constriction is associated with atrial fibrillation (<20%), an early third heart sound and pericardial calcification (25%)
• Diagnosis is on clinical, radiological and echocardiographic grounds. The pericardial effusion is bloodstained in 85% of cases.

Central nervous system disease

• By far the most important form of central nervous system tuberculosis is meningeal disease
• This may complicate primary or postprimary tuberculosis. It can be life-threatening and rapidly fatal if not diagnosed early. Accompanying pulmonary disease (usually miliary) is not infrequent
• Onset is insidious with vomiting, headache, fevers and night sweats occurring over 1–2 weeks
• Mental and personality changes follow with

progressive drowsiness, meningism, cranial nerve palsies (particularly 3rd and 6th), focal long-tract signs and eventually coma in the 3rd to 4th week. Hyponatraemia is common
• Lumbar puncture reveals a lymphocytic cellular response (10–400/μL) with raised protein and depressed CSF : serum glucose ratio
• Tuberculomata are uncommon and usually present with focal features.

Bone and joint disease

• Any bone or joint can be affected but the commonest are the spine and the hip
• Tuberculosis of the spine usually presents with chronic back pain and typically involves the lower thoracic and lumbar spine. Disc involvement is the first feature followed by spread along the spinal ligaments to involve adjacent anterior vertebral bodies causing angulation of the vertebrae and kyphosis. Paravertebral and psoas abscess formation are not uncommon
• Computed tomography scanning helps gauge the extent of disease, the amount of cord compression, and the site for needle biopsy or open exploration if required. The major differential diagnosis is malignancy, which tends to affect the vertebral body and leave the disc intact
• In the absence of spinal instability or cord compression, patients can be treated as outpatients
• Presentation of joint disease is insidious with pain and swelling; fever and night sweats are uncommon. Radiological changes are often non-specific but as disease progresses, reduction in joint space and erosions appear.

Genitourinary disease

• Renal disease is uncommon and is often very insidious with minimal constitutional symptoms
• Haematuria, frequency and dysuria are often present, with sterile pyuria found on urine microscopy and culture
• In men, genitourinary tuberculosis may present as epididymitis or prostatitis
• In women, infertility from endometritis, or pelvic pain and swelling from salpingitis or a tubo-ovarian abscess may occur infrequently.

Diagnosis

• See Table 14.4
• Mycobacterial infection can be confirmed by direct microscopy of samples (positive in 60% of pulmonary and 5–25% of extrapulmonary cases) and culture
• Confirmation that the isolate is *M. tuberculosis* is made through standard culture methods (growth characteristics, pigment production and biochemical tests) or molecular DNA technology (hybridization probes, PCR amplification)
• Drug susceptibility profiles can be obtained within 1–2 weeks of growth using the BACTEC system. Where MDRTB is suspected, molecular methods allow the detection of rifampicin resistance (a marker for multidrug resistance) in cultures as well as primary sputum specimens
• If a cluster of cases suggests a common source, fingerprinting of isolates with restriction fragment length polymorphism (RFLP) or DNA amplification can help confirm this
• Primary tuberculosis in children is rarely confirmed by culture
• In 10–20% of patients with pulmonary and 40–50% of patients with extrapulmonary disease, culture is also negative and the diagnosis is clinical.

Treatment

• Standard therapy consists of 2 months of four drugs (rifampicin, isoniazid, pyrazinamide, and ethambutol) followed by 4 months of rifampicin and isoniazid. It is recommended for all patients with new-onset uncomplicated pulmonary or extrapulmonary tuberculosis. Drugs should be given as a single daily dose before breakfast. Combined drug preparations (including rifampicin and isoniazid with or without pyrazinamide) reduce tablet load and allow relatively simple screening for compliance as the urine can be assessed visually for an orange-pink colour
• Streptomycin is now rarely used in the UK but

Clinical	History of foreign residence, recent immigration or HIV infection
	Subacute respiratory illness with fever
	Chest X-ray showing a cavity or bilateral disease (Fig. 14.1)
Standard tests	Raised ESR, CRP or plasma viscosity
	Normocytic normochromic anaemia
	Mildly deranged liver function tests
	Tuberculin skin test[6]
Specimen	*Diagnostic tests*
Respiratory	Microscopy stain
Sputum[1,2]	Ziehl–Neelsen
Gastric washing[1,3]	Auramine fluorescence
Bronchoalveolar lavage	Nucleic acid amplification (PCR)
Transbronchial biopsy	Culture
Non-respiratory	Solid (Lowenstein–Jensen, Middlebrook)
Fluid examination[4]	Liquid (e.g. BACTEC)
Tissue biopsy[5]	Empirical treatment[7]

1 3 × early morning samples.
2 Induced with nebulized hypertonic saline if not expectorating.
3 Mainly used for children.
4 Cerebrospinal, ascitic, pleural, pericardial, joint, blood.
5 From affected site. Also bone marrow/liver may be diagnostic in patients with disseminated disease.
6 Low sensitivity/specificity: useful only in primary or deep-seated infection.
7 Response to antituberculous drugs usually seen after 5–10 days.

Table 14.4 Diagnosis of tuberculosis.

is an important component of short-course treatment regimens in developing nations
• In patients with a history of past treatment, four drugs must be used until the sensitivity results are obtained. In the UK, drug resistance in newly diagnosed patients is uncommon (overall <5%) and is more frequently observed in isolates from ethnic minority patients
• Patients should be given 9–12 months' treatment where meningeal disease is present, where there is HIV coinfection, or where drug intolerance occurs and a second-line agent is substituted
• Most patients can be treated at home. Treatment should be supervised as closely as possible for the first 2 weeks, partly to see that compliance is satisfactory and partly to be alert to drug reactions. After the first 2 weeks of treatment, patients can be regarded as non-infectious

• To improve adherence in patients unlikely to comply (e.g. alcoholics), directly observed therapy (DOT) three times a week using higher doses of the first-line agents is often used. In developing nations, DOT also dispenses with the need for initial hospitalization to receive streptomycin, is cost-effective and is less disruptive to patients' lives
• Corticosteroids are beneficial in pericarditis, pleural disease and meningitis, and probably in severe pulmonary disease
• The most important drug reactions are hepatitis (isoniazid, rifampicin, pyrazinamide), optic neuritis (ethambutol), psychosis (isoniazid), peripheral neuropathy (isoniazid) and skin rashes (streptomycin and thiacetazone in HIV patients). They occur in ~10%
• Surgery is still occasionally required (e.g. massive haemoptysis, loculated empyema, constric-

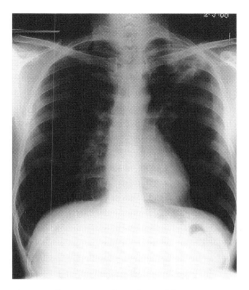

Fig. 14.1 Chest X-ray of a patient with smear-positive pulmonary tuberculosis showing cavity in the left upper lobe.

tive pericarditis, lymph node suppuration, spinal disease with cord compression)
• Relapse is rare when the strain is fully sensitive (<2%) and adherence to the drugs is complete
• Occasionally, paradoxical worsening of disease can occur after commencement of treatment, especially with lymph node disease
• The treatment of MDRTB is complex and depends on the sensitivity of the isolate. Five or more drugs are used and the patient must be admitted to a negative-pressure isolation room for treatment until deemed non-infectious.

Prevention

• The best protection against tuberculosis is the efficient diagnosis and treatment of people with active infections. Tuberculosis of all forms is a notifiable disease in the UK

• Close contacts of patients with pulmonary disease should have their bacille Calmette–Guérin (BCG) and clinical status reviewed, tuberculin skin test (usually Heaf) performed and the need for radiography assessed. The aim of contact tracing is to identify a possible index case with clinical disease, other cases infected by the same index patient (with or without evidence of disease), and close contacts who should receive BCG
• Intradermal tuberculin skin testing is usually performed using the Heaf or Mantoux technique. The response is graded (e.g. Heaf grades 0–4) according to the degree of induration. The test is used to assess whether a person has acquired *M. tuberculosis* following exposure, and is useful in persons not immunized with BCG. It is also used preimmunization with BCG to judge whether or not persons have had previous subclinical primary tuberculosis. Interpretation is more difficult in BCG-vaccinated persons since a mild positive reaction is to be expected
• Chemoprophylaxis is given to prevent infection progressing to clinical disease. It is recommended for children aged <16 years with strongly positive Heaf tests, for children aged <2 years in close contact with smear-positive pulmonary disease, for those in whom recent tuberculin conversion has been confirmed, and for babies of mothers with pulmonary tuberculosis. It should be considered for HIV-infected close contacts of a patient with smear-positive disease. Rifampicin and isoniazid for 3 months, or isoniazid for 6 months are all effective
• BCG is used in some countries as a protective measure for mycobacterial infections. It gives approximately 80% protection for 10–15 years and is greatest for preventing disseminated disease in children. Because of its potential diagnostic value as a measure of recent primary infection, some countries do not use it
• Occasional complications include local BCG abscesses, and disseminated BCG infection in immunocompromised persons.

Prognosis

• With short-course therapy using four first-line drugs, cure is to be expected
• Occasional patients die of overwhelming infection (usually miliary disease, meningitis or bronchopneumonia) and some patients succumb to the later complications of tuberculosis (e.g. cor pulmonale)
• In HIV-associated tuberculosis, mortality is increased, but mainly as a result of superimposed bacterial infections.

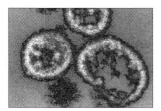

Systemic Mycosis

Fungi are saprophytic organisms widely present in nature (soil and plants); some live on human skin or mouth (e.g. *Candida*). Of the 50 000 or so known species, several are pathogenic to man and may cause a wide range of diseases, from superficial involvement of skin, hair or nails (e.g. ringworm, pityriosporiasis) or mucous membrane (e.g. *Candida*), to involvement of subcutaneous tissue (e.g. mycetoma), to localized or disseminated invasion of the body (systemic mycosis).

Important examples of systemic mycosis are candidiasis, cryptococcosis, aspergillosis, histoplasmosis and coccidioidomycosis, which are dealt with in this chapter. Rarer forms of systemic mycosis (e.g. mucormycosis, fusariosis, systemic sporotrichosis and systemic phaeohyphomycosis) are beyond the scope of this book. However, though previously rare, *Penicillium marneffeii* has emerged as an important opportunistic infection in patients with HIV-associated immunosuppression in Thailand and other parts of south-east Asia, causing disseminated disease presenting with fever, generalized lymphadenopathy, enlarged liver and spleen and multiple papular eruptions with central umbilication.

Candidiasis

Epidemiology and pathogenesis
• The commonest of all fungal infections; most are superficial in nature involving skin or mucous membrane

• Most infections are due to *Candida albicans* which is a commensal of the human mouth and intestine

• Disturbances to the body's intricate defence mechanisms underlie all infections (candidiasis is a disease of the 'diseased') but often remain unexplained

• The severity of mucocutaneous involvement depends on the degree of cellular immunosuppression, whereas systemic invasion usually occurs in neutropenic patients or in debilitated ill patients with infected venous access lines

• Less pathogenic species like *C. tropicalis*, *C. glabrata* or *C. parapsilosis* may cause disease in severely immunocompromised patients.

Clinical features
These depend on the predisposing factors and sites of involvement (Table 15.1).

Treatment
• *Oropharyngeal candidiasis*:
   • mild cases respond to topical nystatin, miconazole or amphotericin B
   • severely immunocompromised patients require an orally absorbable azole drug: fluconazole or itraconazole. Relapses are common and are best managed by intermittent self-therapy
   • azole resistance can become a problem in AIDS patients after recurrent treatments,

Table 15.1 Forms of candidiasis.

Condition	Clinical features	Predisposing factors	Diagnosis
Oral			
Pseudo membranous	White patches on buccal mucosa can be readily wiped off leaving a red, raw surface	Infancy, old age Debility, cellular immune deficiency Steroid inhalation	Clinical Demonstration of *Candida* in scraping or smear if necessary
Acute atrophic	Sore mouth with reddened mucosa; tongue smooth, shiny, may be swollen	Oral broad-spectrum antibiotics	As above
Chronic atrophic	Chronic erythema and oedema of mucosa under the dentures; angular cheilitis may be present	Dentures Chronic mucocutaneous candidiasis*	As above
Oesophageal	Dysphagia May be silent	Severe cellular immune deficiency as in AIDS	Endoscopy Barium swallow
Vaginal	Vulvovaginal pruritus Local discomfort and thick curdy discharge White mucosal patches Vulval erythema Recurrences common	Common in apparently healthy women Pregnancy Diabetes mellitus Tight insulating clothes Antibiotics Cellular immune deficiency	Clinical, smear and culture
Penile	Soreness or irritation of glans penis Discharge, erythema or raised red patches	Intercourse with a person with vaginal candidiasis Diabetes	Clinical, smear and culture
Cutaneous	Erythema with tiny vesiculopustules Groin and submammary areas are commonly affected	Warm, moist skin Friction and occlusion Napkin dermatitis Chronic mucocutaneous candidiasis*	Clinical, smear and culture
Nail and nailfold	Swollen erythematous painful nailfold. Involvement of nail plate may cause discolouration, pitting, friability and separation Bacterial superinfection is common	Repeated immersion in water Chronic mucocutaneous candidiasis*	Clinical, smear and culture
Septicaemic	Fever, metastatic infections in liver, spleen, kidneys, brain, heart, bones	Debilitated patients Indwelling venous lines Neutropenia Intravenous drug use	Blood culture Tissue biopsy

* Chronic mucocutaneous candidiasis is a disease of unknown aetiology with congenitally impaired cellular immunity to *Candida*.

necessitating maintenance intravenous (IV) amphotericin B. The new semisynthetic lipopeptide echinocandins (e.g. casofungin) are probably equally effective, work against azole-resistant strains and are well tolerated (no oral formulation)

• *Vaginal candidiasis:* single-dose flucona-zole (drug of choice) or a topical imidazole (clotrimazole, miconazole or econazole) for 5 days

• *Cutaneous and nail infections:* topical antifungal is useful. Chronic infection requires prolonged oral therapy with daily soaking of the nails using potassium permanganate or phenylmercuric borate solution

• *Systemic candidiasis:* IV amphotericin B with or without oral flucytosine is the drug of choice. Infected lines must be removed.

Aspergillosis

Epidemiology and pathogenesis
• The causative organisms *Aspergillus fumigatus*, *A. niger* and *A. flavus* are commonly found in soil, dust and decaying vegetable matter worldwide
• Infection is via inhalation of airborne spores. Cases are usually sporadic but clusters of cases may occur in cancer wards from environmental sources
• Healthy persons have a high degree of resis-tance; disease susceptibility is increased by:
   • presence of asthma or allergy to *Aspergillus* (allergic bronchopulmonary aspergillosis)
   • bronchitis or bronchiectasis (fungus ball or aspergilloma)
   • immunosuppressive or cytotoxic therapy (pulmonary and disseminated infections)
• Blood vessel involvement with thrombosis and infarction is characteristic of pulmonary and disseminated aspergillosis.

Clinical features
Despite the ubiquity of the organisms, human illnesses are rare. Several forms are recognized.

Allergic bronchopulmonary aspergillosis
• In atopic subjects colonization and prolifera-

tion of *A. fumigatus* in the bronchial passages may lead to allergy to *Aspergillus*, causing cough and wheezy attacks
• Mucosal damage may produce chronic obstructive airways disease and bronchiectasis
• Transient lung infiltrates are visible on chest X-ray (CXR) due to bronchial plugging
• Eosinophilic and serum precipitating anti-bodies are present. Specific immunoglobulin E (IgE) antibody to *Aspergillus* may be demonstrable
• Treatment is as in asthma. Antifungal drugs are of little value.

Aspergilloma
• A pre-existing lung cavity may become colo-nized, leading to formation of a ball of fungal mycelium
• Often asymptomatic, less commonly cough and malaise; sometimes severe haemoptysis may develop
• CXR shows a characteristic round shadow within a cavity with a halo round it
• *Aspergillus* is found in sputum and serum pre-cipitin test is positive
• Asymptomatic and mildly symptomatic pa-tients are only observed, others may need surgi-cal removal. Local instillation of antifungal drugs may help.

Invasive aspergillosis
• In immunocompromised patients *Aspergillus* may invade lung parenchyma, central nervous system (CNS), kidneys, bones and other organs
• Risk is highest in allogenic haematopoietic stem cell transplants followed by neutropenic haematological cancers, autologous stem cell transplants and aplastic anaemia followed by solid organ transplants, AIDS and high-dose corticosteroids
• The condition has a high risk of fatality (up to 85%). Voriconazole may be superior to ampho-tericin B and is better tolerated. Relapse rate is high and secondary antifungal prophylaxis im-proves prognosis
• If feasible, immunosuppressive therapy should be discontinued and neutropenia corrected

- In an outbreak situation, epidemiological investigations to locate the source of spores are necessary.

Ear and nasal sinus infections
- Paranasal *Aspergillus* granuloma:
  - commoner in tropical countries
  - patients often have a long history of ' chronic sinusitis' but are otherwise healthy
  - usually presents with unilateral proptosis, may be painless
  - focal cerebral lesions or cranial neuropathy may develop from CNS invasion
  - CT or MR scan helps in identifying the mass, and histology of excised tissue shows a fibrosing granuloma with occasional hyphae
  - early surgical removal may be curative but CNS involvement is usually fatal
- Superinfection of a chronically infected sinus is commoner, may contribute to symptoms due to allergy, or may form a fungal ball without tissue invasion. Surgery is usually necessary
- Otomycosis—*Aspergillus* is commonly involved.

Histoplasmosis

Epidemiology
- The causative organism *Histoplasma capsulatum* is widely found in soil in eastern and central USA and Africa, and less commonly in the rest of the Americas and south and south-east Asia. It is rare in Europe
- Infections occur in humans, animals and birds
- Bird droppings and decaying organic matter in soil are prominent sources of infection
- Infection results from inhalation of airborne spores.

Clinical features
- Infection is usually asymptomatic
- Less commonly, the patient develops an acute respiratory illness:
  - cough, malaise and fever
  - CXR may show diffuse small opacities
  - the disease is usually self-limiting and mild

in the immunocompetent, rarely chronic disease resembling pulmonary tuberculosis develops
- Disseminated disease may develop in the immunocompromised, e.g. AIDS patients. Presentation is subacute, with fever, weight loss, mild cough and hepatosplenomegaly. Miliary mottling may be present on CXR. Illness may result from reactivation of dormant infection.

Diagnosis
- Serology is diagnostic in the immunocompetent but unreliable in immunocompromised patients
- Demonstration of the organism in bone marrow, blood, sputum or splenic aspirate is confirmatory.

Treatment
- Treatment is not necessary in the immunocompetent patient, who has usually recovered by the time of serological diagnosis
- In disseminated disease, amphotericin B is the treatment of choice although fluconazole and itraconazole are also useful. Maintenance therapy is essential in AIDS patients.

Coccidioidomycosis

- This is caused by the fungus *Coccidioides immitis*
- The organisms are only found in the dry desert areas of western USA, and parts of South and Central America
- Transmission is via inhalation of airborne spores
- Infection is usually asymptomatic or may result in a self-limiting, acute, influenza-like illness
- Disseminated disease may develop in the immunocompetent but is seen more commonly in the immunocompromised such as AIDS patients. Previously asymptomatic dormant infection may activate in such patients
- Amphotericin B is the drug of choice, although fluconazole is proving useful. Recurrences are common in disseminated infection and maintenance therapy is necessary.

Cryptococcosis

Epidemiology

- The causative fungus *Cryptococcus neoformans* is present widely in soil and pigeon droppings
- Occurs worldwide
- It was a rare infection in the pre-HIV era and was seen mostly in patients receiving immuno-suppression therapy. Immunocompetent individuals normally have high resistance to clinical infections, but cases have occurred in patients without demonstrable immunodeficiency
- In immunosuppressed HIV patients, it is a major cause of life-threatening infection
- Infection occurs through inhalation of air-borne spores.

Clinical features

- Meningitis is the usual manifestation:
  - onset is insidious with worsening headache over many days
  - in AIDS patients the symptoms and signs of meningeal irritation may be minimal or absent
- Widespread systemic dissemination may occur:
  - oral, pleural, pulmonary, myocardial, mediastinal, glandular and cutaneous cryptococcosis may be the presenting problem or
  - the patient may present only with fever due to cryptococcaemia.

Diagnosis and treatment

See p. 180.

Infection in Special Groups, Zoonoses, Tropical Diseases and Helminths

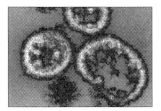

Infection in Special Groups

Pregnant women and neonates

Some infections in pregnancy, during birth or in the neonatal period can be much more severe than at other times. These have been described in other chapters, and this list only summarizes the important ones.

Infections during pregnancy or childbirth
See Table 16.1.

Infections acquired in the neonatal period
• Newborn babies are colonized by whatever microorganisms are in their environment and distributed by their carers
• Neonates, especially the premature, are susceptible to invasion by environmental microbes because they:

• lack antibodies (except passively transferred maternal antibodies)
• lack protective flora
• often have thin skins, raw umbilical stumps, immature inflammatory response
• may be compromised by openings offered by congenital malformations or medical interventions
• Thus skin infections, meningitis, gastroenteritis, epidemic viral diseases and septicaemias are prone to occur
• Outbreaks may occur due to staphylococci, Gram-negative bacilli, and coxsackie- and echoviruses in neonatal nurseries
• Breast feeding is an important route of mother–child human immunodeficiency virus (HIV) transmission
• Chickenpox during the first 28 days tends to be severe (whether contracted from mother or from a carer).

Table 16.1 Infections during pregnancy or childbirth.

Pathogen	Disease more severe if infection occurs during pregnancy?	Effect on fetus or newborn
Rubella	No	Congenital rubella syndrome (early pregnancy infection)
Varicella	Severe rash and pneumonia more likely	Congenital varicella syndrome (early pregnancy infection)
		Herpes zoster during childhood (mid-/late pregnancy infection)
		Severe neonatal chickenpox (rash within 1 week before delivery)
Toxoplasmosis	No	Miscarriage (early pregnancy infection)
		Stillborn or born with chorioretinitis, cerebral calcification and brain damage, hepatitis (late pregnancy infection)
Parvovirus B12	No	Fetal death and abortion, or fetal anaemia
		Congenital malformations do not occur
Syphilis	No	Congenital syphilis
Cytomegalovirus	No	In primary infections, 30–40% of fetuses are infected, of whom 10% are clinically affected (congenital CMV disease)*
		Reactivated maternal infection is generally harmless
Herpes simplex virus	No	Disseminated disease in the newborn if mother gets primary genital herpes during late pregnancy
		Recurrent infection—usually harmless
Hepatitis B	Severe disease more likely	Transmission of infection to baby at the time of birth, followed by chronic carrier state and disease in later life
HIV	No	Transmission to baby usually during birth, less commonly intrauterine—congenital HIV disease
Group B Streptococcus	Urinary tract infections (2–4% incidence) Uncommon—puerperal sepsis, septicaemia	Invasive neonatal disease of early-onset type
Listeria monocytogenes	Meningitis more likely	Neonatal septicaemia or meningitis from transmission during birth or in utero
Chlamydia trachomatis	Postpartum endometritis may develop	Conjunctivitis and pneumonia

*Diagnosis requires isolation of the virus in urine, presence of a compatible clinical picture and exclusion of other causes of similar presentation (HSV, Toxoplasma and rubella infections).

Infections in neutropenic patients and transplant recipients

Neutropenic patients

Neutrophils are an integral part of the host's defence system, responsible for phagocytosis and killing of many invading bacteria and fungi, and for producing the neutrophilic inflammatory response seen typically at the site of invasion. Neutropenic patients are thus vulnerable to a wide range of bacterial and fungal infections.

Pathogenesis

- Risk of infection begins to increase when the neutrophil count drops below $0.5 \times 10^9/L$ and is high at levels below $0.1 \times 10^9/L$
- Risk is higher in acute neutropenia than when neutropenia develops slowly and stabilizes
- At levels $<0.5 \times 10^9/L$ control of mouth flora becomes inefficient and at levels $<0.2 \times 10^9/L$ inflammatory response disappears
- Organisms involved are:
  - usually those commonly found on normal skin, in the mouth and bowel:
    - *Staphylococcus epidermidis, Staphylococcus aureus, Streptococcus viridans, Streptococcus pneumoniae*
    - *Escherichia coli, Pseudomonas* spp., *Klebsiella*
    - *Candida*
  - less commonly, exogenous sources: *Aspergillus, Mucor*
- Indwelling intravenous lines are prone to become colonized with *Candida* and staphylococci.

Clinical presentation

- Mouth ulcers with grey slough, gingivitis and other periodontal diseases, sinusitis
- Anorectal symptoms
- Bacterial pneumonia—sputum is non-purulent and signs of consolidation less clear cut
- Cough, haemoptysis and localized lung infiltrate with central translucency or nasopharyngeal symptoms with haemorrhagic discharge may suggest invasive aspergillosis

- Septicaemia
- Necrotic bullae or pustules (ecthyma gangrenosum), commonly in axillae and groins in *Pseudomonas* septicaemia
- Fever alone is often the presenting feature.

Investigations

- Thorough examination of skin, mouth and nasopharynx, anorectal region, lungs and abdomen for a source of infection
- Several blood cultures, to be incubated for at least 2 weeks (for slow-growing organisms)
- Urine and sputum examination
- Culture of indwelling catheter
- Chest X-ray
- Computed tomography scan of lung or nasopharynx, bronchoalveolar lavage or nasopharyngeal biopsy for diagnosis of invasive aspergillosis.

Treatment

- Appropriate antimicrobial therapy, and surgery, if necessary, for identifiable local infections
- Bactericidal antibiotics should be used whenever possible
- In febrile patients with no identifiable site of infection start empirical broad-spectrum antibiotic therapy to cover Gram-negative and Gram-positive infections (e.g. a third-generation cephalosporin alone (ceftazidime) or a broad-spectrum penicillin plus aminoglycoside)
- If fever is still present on day 7 and cultures are negative add amphotericin B
- Consider granulocyte colony stimulating factor (G-CSF) to stimulate neutrophil production
- Scope of neutrophil transfusion very limited.

Transplant recipients
Pathogenesis

- Transplant recipients develop severe cellular immune deficiency from drugs used to prevent rejection of the transplanted organ. This makes them vulnerable to infections from a large group of pathogens against whom the host defence relies on an intact cellular immunity for prevention of invasion and for recovery:
  - viruses: herpes simplex virus (HSV), vari-

cella zoster virus (VZV), cytomegalovirus CMV), Epstein–Barr virus
- bacteria: Mycobacterium tuberculosis, Listeria monocytogenes, Nocardia
- fungi: Candida, Pneumocystis carinii, cryptococci, Histoplasma, Coccidioides, Aspergillus
- parasites: Toxoplasma gondii
- Risk from these infections varies according to the degree of immunosuppression:
    - immune suppression begins to intensify towards the end of the month following transplant (early period — infection risk low)
    - suppression is intense between 1 and 6 months (middle period — risk high), after which it begins to improve (late period >6 months — risk low)
    - immunity is restored by the end of the year in bone marrow transplant (BMT) recipients but low-level suppression persists indefinitely in solid organ transplant recipients because of continuing need to use immunosuppressants
- During the early period, transplant patients are also vulnerable to pathogens against whom host immunity is not cellular; the pattern of these infections depends on the nature of the transplant.
    - BMT recipients are severely neutropenic during the early period and are prone to the infections discussed earlier
    - BMT patients are also at higher risk from pneumonia due to Streptococcus pneumoniae and Haemophilus influenzae as the immature bone marrow is unable to produce protective antibodies against capsular polysaccharides
    - Solid organ transplant recipients are also prone to develop infectious complications in relation to operation:
        - kidney — urinary tract infection
        - liver — peritonitis, abdominal abscesses
        - heart — mediastinitis.

Prevention of infection in transplant recipients
- HSV,VZV — acyclovir (in BMT)
- CMV — CMV immunoglobulin, ganciclovir in some settings, use of CMV-negative blood in seronegative recipients

- P. carinii, T. gondii, Nocardia — trimethoprim/ sulfamethoxazole
- S. pneumoniae, H. influenzae, influenza — vaccination.

Patients with pyrexia of unknown origin (PUO)

Clinicians often use the term PUO to characterize persistent fevers for which a cause has not been found even after adequate history-taking, detailed and multiple physical examinations and an initial phase of investigations, preferably in hospital settings.

The qualifying duration of fever varies according to clinical settings.
- For community-acquired fever in a previously healthy individual it usually involves a duration of 2 weeks, during which:
    - the majority of common viral infections which often present only as fever will have settled
    - other infections which initially present with a fever prodrome will have developed specific diagnostic features (e.g. viral hepatitis, exanthemata)
    - the diagnosis of malaria, most cases of typhoid presenting in the West, septicaemias, clinically silent pneumonias, and most pulmonary and haematological malignancies should have become evident
- For neutropenic and immunosuppressed individuals (e.g. HIV patients and transplant recipients) the term is applicable once blood cultures and other baseline investigations prove negative even if fever duration is shorter.

Causes of PUO
Infections
Community-acquired PUO
Modern diagnostic techniques have significantly reduced the number of infection-related cases to < 20%.
- Tuberculosis remains an important cause, particularly in immigrants from the Indian subcontinent and Africa in whom the disease is often

extrapulmonary, and diagnosis may be difficult in the following clinical types:
- cryptic or disseminated (miliary opacities not detectable on chest X-ray (CXR))
- mediastinal lymphadenitis (widened mediastinum not apparent on CXR)
- abdominal tuberculosis (fever and vague abdominal discomfort only)
- *Hidden abscesses* — commonly intra-abdominal (e.g. subphrenic, intrahepatic, pancreatic, renal, splenic, paracolic, paraspinal and pelvic)
- *Chronic prostatitis and pelvic sepsis* are important causes of recurring fever
- *Endocarditis* — significance of a seemingly innocent murmur not realized and blood cultures are negative from prior antibiotics or organisms are slow-growing (*Haemophilus* spp., *Actinobacillus actinomycetemcomitans*, *Eikenella* and *Kingella* spp.) or patient has Q fever endocarditis
- *Glandular fever* — throat involvement may be minimal and Monospot test initially negative
- Undiagnosed *HIV infection*
- *CMV* and *Coxiella burnetii* may cause subacute granulomatous hepatitis, presenting only with fever and low-grade liver function abnormalities
- *Amoebic hepatitis* — ultrasound may be negative in the early stages
- *Brucellosis, visceral leishmaniasis and trypanosomiasis.*

Neutropenic patients
Fever is usually infection related, mostly bacterial and sometimes fungal, but blood cultures are often sterile (see p. 205).

Immunocompromised HIV patients and transplant recipients
An infective cause is much more likely.
- Tuberculosis (often extrapulmonary, or CXR changes may be less developed)
- *Mycobacterium avium* complex — a common cause of PUO in HIV patients with $<50\,mm^3$ CD4
- Cryptococcal meningitis — meningeal signs often absent or muted
- *Pneumocystis carinii* pneumonia — CXR may look normal initially or infection may be extrapulmonary

- HIV infection itself.

Neoplasms
- Lymphomas are an important cause of PUO — in both previously healthy and HIV patients
- Hypernephroma, metastatic carcinoma and intra-abdominal malignancies.

Connective tissue disorders
Systemic lupus erythematosus, polyarteritis nodosa, Still's disease, mixed connective tissue disease — can manifest only with fever.

Miscellaneous
- Sarcoidosis
- Kawasaki disease
- Whipple's disease
- Recurrent multiple pulmonary embolisms of pelvic origin
- Crohn's disease
- Familial Mediterranean fever
- Hyperimmunoglobulin D syndrome

Diagnostic approach
Only a list of usually helpful investigations is given here — in practice these have to be tailored to individual circumstances.
- Thorough reappraisal of clinical presentation, travel history, occupational exposure
- Repeated and thorough clinical examinations for any localizing sign
- Observation of fever pattern and ensuring that the fever is not spurious (patient's own record, general condition good/pulse does not rise with fever/plasma viscosity and C-reactive protein (CRP) not raised)
- Review of all available laboratory results for any abnormality and repeat base line tests
- Significant neutrophilic leucocytosis and raised plasma viscosity and CRP make a pyogenic infection more likely
- Search for an infective cause (longer the duration of fever (>1 month) — the less likely an infective aetiology in community-acquired cases):
    - further blood cultures — think of fungi and slow-growing organisms
    - repeat Monospot test

• serological tests — as the patient has been ill for more than 2 weeks, collect serum for evidence of antibody response against *Coxiella*, CMV, *Mycoplasma*
• culture of induced sputum or morning gastric aspirate and urine in suspected disseminated tuberculosis (TB)
• in the immunocompromised look for TB and MAC in bronchoalveolar lavage fluid and bone marrow aspirate, cryptococcal and CMV early antigen in blood
• HIV test following appropriate counselling if patient has unexplained lymphopenia/thrombocytopenia, or history of recurrent oral thrush or herpes zoster (in young)
• repeat abdominal ultrasound if earlier one negative but amoebic hepatitis still likely
• search for hidden abdominal abscesses or malignancy by ultrasound, CT or MR scan
• CT lung scan helps in identifying early miliary TB or *Pneumocystis carinii* pneumonia (PCP) even when CXR is normal
• gallium scan is helpful in identifying an inflammatory focus but does not distinguish between infective and non-infective causes. Gadolinium MR scans are better for locating inflammatory lesions
• Connective tissue disease screen
• Biopsy of:
  • liver may help if there is any hepatic dysfunction, however minimal (i.e. in sarcoidosis, granulomatous hepatitis)
  • any enlarged gland or skin lesion
  • peritoneum (TB peritonitis)
  • bone marrow (disseminated TB, MAC, typhoid, visceral leishmaniasis, infiltrative neoplasms)
  • mediastinal glands (TB, sarcoid, lymphoma)
  • any deep-seated lesion (guided by ultrasound)
  • intestine (Crohn's, TB, Whipple's disease).
Occasionally, even after exhaustive investigations, a cause is not found. In some of these cases fever settles spontaneously without a diagnosis ever being achieved, in others, the nature of the disease becomes apparent only in time.

Infection in travellers

Visit to a distant land often brings travellers in contact with infections not prevalent in their homelands and thus puts them potentially at risk.

Activities such as contact with large groups of people during transit and stay, eating and drinking, contact with insects and animals, and recreational and sexual pastimes create opportunities for pathogens to be transmitted.

The following text summarizes the common or important travel-related infections according to their modes of transmission. They are discussed in detail elsewhere.

RISK FACTORS FOR TRAVEL-RELATED INFECTIONS

Travel-related infections are commoner in residents of industrialized nations travelling to resource-poor developing countries who:
• travel on a low budget and stay in crowded conditions
• live close to local people for prolonged periods (e.g. back packers, immigrants (or their children) visiting relatives in their countries of origin)
• have a casual attitude to food and drink safety
• indulge in unprotected casual sex.

Infections acquired through contact with other people

Transmission is either airborne or via bodily contact. The risk varies according to the closeness, duration and nature of the contact.

Casual contact in public places

• Only the highly infectious pathogens can transmit in such circumstances — e.g. influenza and other respiratory viruses, *Mycoplasma*, Norwalk-like viruses (NLVs)
• Respiratory viral infections are common among travellers
• These occur worldwide but seasonal variations are common (e.g. influenza season in Northern hemisphere is between December

and March, but between April and September in Southern hemisphere)

• Outbreaks of respiratory virus infections are common when people are crowded together in hotels and clubs at tourist resorts, at pilgrimage sites, and in buses and trains

• Outbreaks of NLV-related diarrhoeas are common in hotels, camp sites and cruise ships when the virus spreads easily from person to person.

Infections that require more prolonged contact in confined places

• Meningococcal infections—epidemics and outbreaks are common in tropical sub-Saharan Africa, the Indian subcontinent and the Middle East. Travellers living among local people in crowded conditions are at risk and should be vaccinated

• Diphtheria—prevalent in the former Soviet Union countries, the Indian subcontinent and Africa. Non-immunized people working closely with children for prolonged periods (e.g. teachers) are at risk and should be vaccinated

• Tuberculosis—highly endemic throughout the developing world, especially the Indian subcontinent and Africa. Risk to ordinary travellers is remote; at risk are children of immigrants visiting relatives who should have their BCG status checked.

Non-sexual bodily contact

• Streptococcal skin infections and scabies are common in children of poor socioeconomic groups in the tropics, and visiting children staying with relatives are at risk.

Close contact with patients

• Health-care workers and volunteers working closely with patients in hospitals in areas for viral haemorrhagic fever (VHF) are at risk.

Infections acquired through food and drink

Traveller's diarrhoea

• Is the most common ailment affecting travellers from the industrial West to developing countries

• Risk varies according to the countries of travel (survey among European travellers):
  • Indian subcontinent—50%
  • west Africa—40%
  • Peru—34%, Brazil—26%
  • Central America—26%
  • south-east Asia—25%
  • North America—4%

• Causative pathogens vary according to geography; overall 50–70% are bacterial
  • E. coli, mainly enterotoxigenic E. coli—5–70%
  • Campylobacter—0–30%
  • Salmonella—0–15%
  • Shigella—0–15%
  • Aeromonas and Plesiomonas—0–10%
  • Up to 10% of diarrhoea lasting for >2 weeks is caused by a parasite: Giardia, Cryptosporidium, Cyclospora and Isospora. Amoebiasis is rare

• Diarrhoea subsides within 5 days in most cases and stool examination is unnecessary unless symptoms are severe or persist beyond 2 weeks (for parasites)

• Attention to fluid and electrolyte loss is generally all that is necessary

• A single dose of ciprofloxacin is often effective and is an appropriate self-therapy along with loperamide for a day

• Prevention:
  • advice on food and drink safety—drink bottled or boiled water; avoid ice, peeled fruit and salad; eat hot, freshly cooked food; avoid street vendors
  • provide self-therapy
  • use selective chemoprophylaxis with a quinolone—traveller on a very short tour for an important mission
  • bismuth subsalicylate is also effective (less so) and the tablet form may suit someone on a longer visit who is unwilling to stick to food safety.

Typhoid and paratyphoid fevers

Such cases in the developed countries are almost always contracted abroad, mostly in the Indian subcontinent, South America and Africa. Most cases occur in immigrants on long visits to their homelands—the risk to short-term tourists is quite low.

Hepatitis A and E

Hepatitis A is more prevalent in most resource-poor countries and hepatitis E exclusively so, and travellers with casual attitude to food and drink safety are at significant risk (vaccination is highly protective against hepatitis A).

Infections acquired through insect bites

Malaria

The commonest life-threatening travel-related infection. It usually presents in short-term travellers after returning home. It occcurs throughout the tropics and subtropics. Risk of infection is high in persons visiting highly endemic areas and neglecting prophylaxis and insect repellent measures.

Dengue

A common cause of fever among travellers to Central America, India, south-east China and south-east Asia. Tourists staying for longer periods and living in rural areas with poorly screened accommodation are particularly at risk. It is difficult to avoid mosquito bites because of their day-time feeding habit.

Japanese B encephalitis

Endemic in rural areas of south-east Asia, especially wet areas such as paddy fields, where pigs and wading birds are present. Risk to travellers is extremely low, but prolonged rural exposure during the transmission season contributes to the risk—consider vaccination.

Yellow fever

Prevalent in the jungles of Africa and South America but urban disease outbreaks also occur when the usual animal host monkeys stray into towns. Vaccination is necessary for those visiting endemic areas.

West Nile fever

This mosquito-borne arbovirus infection is currently causing an epidemic in the USA and spreading. Visitors are at risk.

Lyme disease and tick-borne encephalitis (TBE)

At risk are travellers to the densely wooded temperate climate forests of Europe, North America and Asia in spring and summer. Vaccination should be considered for TBE.

Infections acquired through soil and water

Cutaneous larva migrans

At risk are those walking barefoot on soil and beaches contaminated with dog or cat faeces containing larvae of animal hookworm. Prevalent in Central and South Americas and Africa. Establishment of human hookworm infection requires repeated infections.

Legionellosis

Aerosols of infected water from air-conditioning, showers or jacuzzis have caused outbreaks in hotels and recreational facilities worldwide.

Bilharziasis

Contracted by swimming in fresh water containing schistosome cerceriae in Africa, South America and the Far East. May present acutely as fever, skin lesions or myelitis within a few weeks of a single water contact. Chronic disease requires prolonged and repeated infections.

Leptospirosis

Prevalent worldwide, occurring through contact of abraded skin or mucous membrane with water and soil contaminated with infected animal urine. A recreational hazard to bathers, campers or sportsmen in infected areas.

Infections acquired through animal contact

Rabies

Endemic in India, China and many south-east Asian countries, mostly from dog bites. Recently bats have caused human infections in Europe. Prompt postexposure prophylaxis is usually effective. Pre-exposure vaccination obviates the need for hyperimmune gammaglobulin and buys time for initiating treatment.

Sexually transmitted infections

• Prevalence of many sexually transmitted diseases is higher in most developing countries
• Unprotected casual sex and sex with prostitutes carry a significant risk, especially of gonorrhoea (often penicillin resistant), HIV and hepatitis B
• In many developed countries, heterosexually acquired gonorrhoea in males is predominantly acquired abroad.

Helpful websites
• www.cdc.gov/travel
• www.hpa.org.uk/infections/topics-az/travel/menu.htm

Infection in injection drug users

The incidence of infection among injection drug users (IDUs) is much higher than among the population in general.
• They have a much higher premature death rate, the majority infection related
• 20–60% of all hospital admissions of IDUs are due to infection.

Pathogenesis
• Most infections are related to use of non-sterile equipment and solutions, and poor skin cleansing
• Infections also occur secondary to HIV-related immunosuppression and due to poor socioeconomic conditions (e.g. tuberculosis).

Microbiology
• Skin flora: staphylococci, streptococci
• Perineum flora: enteric Gram-negative organisms
• Mouth/nasopharyngeal flora (use of saliva): viridans streptococci, staphylococci, anaerobic bacteria
• Contaminated water: faecal flora
• Blood contamination: HIV, hepatitis B, C and D—lifetime attack rate is between 80 and 90%, commonest being hepatitis C
• Unusual organisms—e.g. Clostridium botulinum.

Clinical features (of injection-related bacterial infections)
Localized
• Skin and soft tissue infections—commonest: cellulitis, abscesses, necrotizing fasciitis, pyomyositis
• Septic thrombophlebitis.

Systemic
• Pneumonia
• Septicaemia
• Metastatic abscesses
• Endocarditis (often right sided)
• Wound botulism.

Prevention
• Discontinue injecting—ideal goal but difficult to achieve
• Education as to harm reduction methods that minimize risk of soft tissue and invasive bacterial infections significantly
• Use of clean syringe and needles
• Good skin cleansing
• Needle exchange schemes help reduce transmission of blood-borne viruses
• Screening for sexually transmitted disease, blood-borne viruses and tuberculosis
• Immunization against hepatitis B, pneumonia, influenza, tetanus.

Infection in hospitalized patients

Hospital-acquired (nosocomial) infections have been important for as long as infection has been understood. Their prevalence is 10%, their incidence 5%; they cost lives and money. Urinary tract, wound and respiratory tract infections are the commonest.

Contributory factors are:
• increase in the number of susceptible persons in hospitals (elderly, debilitated or immunosuppressed patients)
• increased use of invasive procedures (e.g. indwelling venous lines, assisted ventilation) or suppression of immune system
• increased patient movement between wards due to pressure on beds

• wider use of antibiotics and emergence of resistant strains.

The causative organisms come from five main sources:

• aerosol (e.g. *Legionella*, VZV)
• faecal–oral (e.g. *Clostridium difficile*, *Salmonella*)
• hand or body contact (e.g. methicillin-resistant *Staphylococcus aureus* (MRSA), Gram-negative bacilli)
• contaminated vehicles (e.g. food—*C. perfringens*, *Salmonella*, or equipment—hepatitis B)
• blood products (e.g. hepatitis B, hepatitis C, HIV).

Organization in infection control

Effective control depends on:

• adherence to basic infection control principles
• existence of:
  • an infection control team
  • regular surveillance of high-risk units (e.g. intensive care)
  • monitoring of laboratory results
  • development of policies on topics such as isolation, disinfection and antibiotic usage.

Principle of isolation

The level of isolation will depend on the microorganism, the site of infection, the susceptibility of others and the medical needs of the patient. With this in mind the following principles should be observed.

• Patients with possible infective diarrhoea should be in a single room under enteric precautions
• High secure isolation using a negative-pressure ventilation system is usually required when the patient poses a risk to other patients and staff (e.g. multidrug-resistant TB, VHF)
• Disposable plastic aprons should be used for all intimate patient care when soiling of clothing is likely, when dealing with body sites where there is a break in the normal defence system (e.g. wound and catheter care), when the patient has a skin disorder and is a likely heavy disperser of organisms, and when the patient is under protective isolation

• Hand-washing should be carried out after patient contact. Health-care workers should have intact skin or should cover cuts with a waterproof dressing. Soap and water is nearly always sufficient for hand cleansing but in high-dependency areas an antiseptic preparation is necessary
• Gloves should be worn when there is possible contact with secretions, excretions or blood, or where gross hand contamination is likely
• Masks, full-length gowns, overshoes and hats are only needed when dealing with a patient who is a heavy disperser of multiresistant organisms (e.g. MRSA) or is in high secure isolation.

Multiresistant infections

These can be innately resistant (e.g. *Xanthomonas maltophilia*) or selected resistant strains through hospital and patient antibiotic use (multiresistant *Klebsiella*).

The most important are:

• multiresistant Gram-negative bacilli (e.g. *Pseudomonas*, *Acinetobacter*)
• MRSA
• vancomycin-resistant enterococci.

MRSA strains are becoming an increasing problem and few hospitals are spared. They show the following characteristics.

• Resistant to all β-lactam antibiotics and, until recently, only reliably sensitive to vancomycin but reports of vancomycin-resistant strains are increasing
• Containable only by a strict policy of isolation and regular screening of the index patient and contacts (staff and other patients). Elimination can usually be achieved by eradication of patient carriage (e.g. mupirocin nasal ointment for nasal carriage) and treatment of any infection, e.g. vancomycin and rifampicin or streptogrammin therapy)
• Particularly problematic in patients with skin disorders who are heavy shedders of colonized skin scales and often the origin of outbreaks.

Vancomycin-resistant enterococci (usually *Enterococcus faecium*) show the following characteristics.

• They are increasingly recognized as causing significant disease in compromised hosts

• They are usually also resistant to ampicillin and aminoglycosides (agents normally active)
• They are being recovered from patients in large health-care centres where cephalosporins and vancomycin use is widespread
• Streptogrammin is effective.

Infections acquired from the hospital environment or equipment

Multiresistant Gram-negative bacilli, *Legionella* species and *C. difficile* are the most important in this category.
• Legionellosis has caused large outbreaks of hospital-acquired pneumonia when it has contaminated the ventilation systems, and sporadic cases have been linked to hospital shower systems. Control policies now exist in hospitals to reduce the risk of such outbreaks occurring and emphasize the need for control of hot water temperature, cleaning and disinfection of water tanks and humidifiers, and avoidance of water-cooled air-conditioning systems
• Multiresistant Gram-negative bacilli mainly result from in-hospital selection pressure through intensive medical and surgical practices and broad-spectrum antibiotic use, and are unavoidable. They survive well in the environment and are readily transmitted between patients. Isolation, hand-washing and standard infection control nursing procedures help to reduce incidents
• *C. difficile* has become a common and important nosocomial infection. Its rise has paralleled the equally swift increase in broad-spectrum antibiotic use which selects out *C. difficile* by interfering with commensal gut flora. Key areas in controlling an outbreak are immediate and effective isolation of all patients with diarrhoea, urgent sample collection and same-day toxin testing, immediate and rigorous cleaning of the patient's environment, and review of the antibiotic policy to reduce broad-spectrum antibiotic use.

Outbreak infections

An outbreak is defined as two or more cases that are linked in time or place and may be of three types:

• the 'point source' in which there is a single exposure that is not repeated (e.g. *C. perfringens* food poisoning)
• the 'continuous source' where the source remains infective (e.g. a surgeon with hepatitis B)
• the 'propagating' source where the original source has resulted in secondary spread (e.g. *Salmonella* infection).

The major organisms that cause hospital outbreaks are:
• enteric pathogens (*Salmonella*, *C. difficile*)
• food-poisoning pathogens (*C. perfringens*, *S. aureus*)
• respiratory pathogens (*Legionella* spp., influenza)
• skin pathogens (*S. aureus*)
• pathogens from contaminated IV preparations or an infectious carrier.

Clinical site of disease

The commonest sites of nosocomial infection are the urinary and respiratory tracts, and skin and subcutaneous tissues. The disease is often innocuous but occasionally may be life-threatening (e.g. septicaemia). Numerically less, but individually more important are infections of prosthetic heart valves, orthopaedic implants, intravenous shunts and intravascular devices and grafts, all of which may be hospital acquired.

Urinary tract infection (UTI)

Hospital-acquired UTI is the commonest nosocomial infection, accounting for up to one-third of all hospital-acquired infections and developing in 2–5% of all admissions: 80% are catheter-associated, 1–3% become bacteraemic with 12% mortality.

Following catheterization:
• the incidence of bacteriuria using closed drainage increases by 5–10% each day and by day 30 it is almost universal
• organisms ascend urethrally, external to the catheter
• initial asymptomatic bacteriuria progresses to symptomatic infection in many patients
• in long-term catheterized patients, up to two-thirds of febrile episodes result from UTI

• the organisms involved are typically 'coliforms' (*E. coli*, *Proteus* spp., etc.) and enterococci; occasionally, *P. aeruginosa* and *Candida*
• many infections cause little illness and disappear when the catheter is removed. However, there is a risk of Gram-negative septicaemia at any time, especially if a bacteriuric patient is recatheterized or undergoes urinary tract surgery.

Skin and subcutaneous tissue infections
• Postoperative wound infections may result from:
  • intraoperative contamination with endogenous flora
  • postoperative inoculation of endogenous flora (e.g. *S. aureus* from nasal carriage)
  • exogenous contamination (environmental or cross-infection)
• The most important of these is intraoperative contamination and its risk depends on the degree of contamination of the operation site:
  • 'clean' operations (no infection encountered, no break in aseptic technique, no colonized viscus opened)—minimal risk (e.g. hernia repair)
  • 'clean contaminated' operations (minimal contamination from opened colonized viscus or break in aseptic technique)—moderate risk (e.g. uncomplicated appendectomy)
  • 'contaminated' operations (significant contamination from colonized opened viscus, acute inflammation without suppuration, or traumatic wounds <4h after injury)—significant risk (e.g. major colorectal surgery)
  • 'dirty' operations (frank pus present, perforated viscus or traumatic wounds >4h after injury)—high risk (e.g. perforated colonic diverticulum with paracolic abscess)
• The organisms recovered reflect the colonization flora of the viscus involved. For bowel surgery, 'coliforms' and *Bacterioides fragilis* are most frequently isolated
• Host (age, obesity, immunosuppression, underlying major illness) and surgical (technical skill) factors are also important determining factors.

Respiratory tract infections
• The incidence of nosocomial pneumonia is 4–8/1000 hospital admissions
• It is higher in comatose patients, those with left ventricular failure and those anaesthetized and intubated (up to 40% in ventilated patients, increasing by 1% for every day of ventilation)
• The organism is likely to be Gram-negative, with 'coliforms', *Pseudomonas aeruginosa* and *Acinetobacter* predominating; enterococci (especially if on broad-spectrum cephalosporins) and *S. aureus* are also seen.

Foreign body-related infections
• These are more serious nosocomial infections, often aided by the presence of a foreign body (e.g. prosthetic joint), when a much lower level of contamination is enough to cause infection
• Infections are characteristically chronic and due to low-grade pathogens (e.g. *Staphylococcus epidermidis*) where production of a glycocalyx enables adherence to the implant
• Distinction from contamination is always a problem: recovery usually necessitates removal of the prosthesis.

Useful websites
• www.epic.tvu.ac.uk
• www.hpa.org.uk/infections/topics_az/hai/menu.htm
• www.doh.gov.uk/hai/index.htm

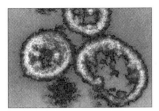

Tropical Infections and Non-helminthic Zoonoses

This chapter deals with infections that are common or important in the tropics and also some of the important zoonotic infections (infections that are transmissible under natural conditions from animals to humans) which have not been dealt with in the previous chapters.

Zoonotic helminthic infections (*Toxocara*, hydatid disease, *Taenia*, *Trichinella*) are dealt with in Chapter 18.

Malaria

Epidemiology
Malaria is caused by parasites of the *Plasmodium* genus and is the most important protozoal infection worldwide.
• 300 million persons contract malaria annually and 1 million die, mostly children <5 years in sub-Saharan Africa
• Two-thirds of reported cases occur in Africa, the Indian subcontinent, Vietnam, the Soloman Islands, Columbia and Brazil
• Between 10000 and 30000 residents of industrialized countries also contract malaria annually through travel to endemic areas
• Around 2000 cases occur annually in the UK with 10–15 deaths (from *P. falciparum*)

• Four human species of malaria exist (*P. ovale*, *P. malariae*, *P. vivax* and *P. falciparum*):
 • *P. vivax* predominates in India, Pakistan, Bangladesh, Sri Lanka and Central America
 • *P. falciparum* is dominant in Africa and New Guinea
 • both are prevalent in south-east Asia, South America and Oceania
 • *P. ovale* and *P. malariae* occur mainly in Africa
• Transmission is through the bite of a female anophelene mosquito; man is the only reservoir
• The mosquito is initially infected by ingesting male and female gametocytes during blood feeding on an infected human. A sexual cycle then takes place in the mosquito with sporozoites appearing in the salivary glands. These are injected during a subsequent feed
 • Sporozoites quickly enter the liver (pre-erythrocytic phase) where they mature and multiply into schizonts which release numerous merozoites to invade red cells, where further maturation and multiplication take place (erythrocytic phase)
 • The erythrocytic cycle lasts for 48 h with *P. falciparum*, *P. vivax* and *P. ovale*, and 72 h with *P. malariae*, following which more merozoites are released which invade fresh cells
 • Materials from disintegrating red cells

activate macrophage and release proinflammatory cytokines that cause the fever
- Eventually merozoite release is synchronized to occur simultaneously, giving rise to the typical periodic fevers (tertian — every 3 days, quartan — every 4 days). Such patterns are rarely seen in the West because of early diagnosis and treatment
- In the case of *P. vivax* and *P. ovale*, the parasites may persist in the liver (hypnozoites), causing later relapses. In *P. falciparum* there are no hypnozoites. *P. malariae* rarely relapses, due to persistent extrahepatic parasites
- The average incubation period (IP) is approximately 12 days for *P. falciparum*, 13 days for *P. vivax*, 17 days for *P. ovale* and 28 days for *P. malariae* although infections may occur up to 3 months after leaving an endemic area for *P. falciparum*, 5 years for *P. vivax* and *P. ovale*, and 20 years for *P. malariae*.

Pathology and pathogenesis

- Parasitized red blood cells are actively removed by the spleen, others are destroyed intravascularly when schizonts rupture; thus splenomegaly and anaemia are common
- In *P. falciparum* infection, by sequestering in and obstructing the capillaries, parasitized cells cause anoxia, lactic acidosis and leaky capillaries. This results in oedema, congestion and microhaemorrhage, which in turn lead to the complications of malaria
- Many cytokines are increased during acute malaria (including tumour necrosis factor-α (TNF-α)) but their exact role remains to be defined
- Raised intracranial pressure has been linked with cerebral malaria but it is uncertain whether this contributes to coma or death
- Immune complex formation (*Plasmodium* antigen, immunoglobulin G (IgG) and complement) in the kidneys of children with *P. malariae* may lead to the nephrotic syndrome
- Tropical splenomegaly syndrome results from an unusual form of immune response to chronic malarial antigenaemia
- Receptors on merozoites and on the red cells are essential for cell invasion

- Sickle cell trait protects against severe *P. falciparum*, as does haemoglobin F; lack of the Duffy blood group antigen protects against *P. vivax*
- Development of specific immune responses eventually controls disease symptoms but asymptomatic parasitaemia occurs commonly, as seen among individuals living in highly endemic areas. The mechanism of immunity is complex and ill-understood but involves both humoral and cellular immunities, which are protective against high parasitaemia and disease. This immunity is lost after some time in individuals living outside endemic areas.

Clinical features

Malaria cannot be diagnosed with confidence clinically, nor can uncomplicated falciparum malaria be distinguished from the other three malarial species. Typically there is:
- an abrupt onset of fever and influenza-like symptoms
- periodicity developing after several days, with episodic fever separated by relatively symptom-free periods — unlikely in primary infections and often does not develop in *P. falciparum infection*; daily fever is common
- episodes of fever consisting of cold, hot and sweating phases, in total lasting about 6–8 h
- anaemia after several days
- hepatosplenomegaly in one-third and jaundice (in *P. falciparum*).

Complications

Severe P. falciparum malaria
Complications are usually associated with high parasitaemia, but severe disease can occur with low parasitaemia since the peripheral load may not reflect the total load (it is the sequestrated mature parasites that cause the tissue damage).

Also, patients initially without complications may deteriorate and parasite count may rise over the next 48 h despite treatment because of the 48-h cycle of maturation and the relative insensitivity of immature parasites (trophozoites) to antimalarial drugs.
- Cerebral malaria — see below
- Severe normochromic anaemia — from haemolysis and bone marrow suppression

- Oliguric renal failure—from acute tubular necrosis
- Pulmonary oedema and adult respiratory distress syndrome
- Hypoglycaemia—quinine-induced hyperinsulinaemia or due to high parasitaemia
- Shock state (algid malaria)—usually due to concomitant Gram-negative septicaemia
- Lactic acidosis
- Spontaneous bleeding, disseminated intravascular coagulopathy (DIC)
- Haemoglobinuria (black-water fever).

Other forms of malaria
- Recurrent infection with *P. vivax*, *P. ovale* and *P. malariae*
- Chronic anaemia
- Ruptured spleen (especially in *P. vivax*)
- Nephrotic syndrome with chronic *P. malariae* infection
- Tropical splenomegaly syndrome.

Cerebral malaria
- Usually develops after several days in adults but in children the history is often < 2 days
- Seizures, increasing obtundation and the development of abnormal neurology (e.g. dysconjugate gaze, decerebrate posturing) develop, sometimes with very rapid progression
- Retinal haemorrhages are common
- Coma may persist for several days after the clearance of parasites from the blood.

Diagnosis

The differential diagnosis is wide, as *P. falciparum* malaria can mimic many diseases (e.g. hepatitis, meningitis, septicaemia, typhoid, arbovirus infection, typhus, gastroenteritis).

Non-specific features supporting a diagnosis of malaria include:
- a consistent clinical and travel history
- splenomegaly
- thrombocytopenia
- abnormal behaviour or mental state (a danger signal: such cases should be treated as a medical emergency).

Confirmation of malaria can be made:
- by thick and thin film microscopy. Up to three

smears should be examined if the history remains suggestive
- alternatively, by antigen detection (kit forms available for field use) or polymerase chain reaction (PCR).

Treatment

Benign relapsing malaria (P. vivax, P. ovale,
P. malariae)
- Chloroquine is the drug of choice for 3 days followed by
- Primaquine for 14 days (to eliminate the hepatic reservoir of infection and prevent relapses) in *P. vivax* and *P. ovale* (in pregnant women wait till pregnancy is over and cover with weekly chloroquine)
- In south-east Asia malaria, primaquine should be given for 3 weeks because of relative resistance
- Levels of the enzyme glucose-6-phosphate dehydrogenase (G6PD) must be checked before primaquine is given, since haemolysis may develop.

Malignant malaria (P. falciparum)
- Quinine for 7 days (with concurrent tetracycline or single-dose Fansidar (day 7)) is the drug of choice
  - Side-effects are to be expected: tinnitus, deafness, nausea and vomiting (cinchonism)
  - Intravenous (IV) quinine should be given if the patient:
    - is vomiting
    - has a high parasitaemia (>4%)
    - has any complication as listed above
  - Patients should be monitored for hypoglycaemia and cardiac arrythmias
- In most parts of the world *P. falciparum* is resistant to chloroquine which should not be used
- Alternative drugs are:
  - mefloquine—resistance reported in many parts of south-east Asia
  - atovaquone/proguanil combination (Malarone)
  - artemisinin and derivatives (artemether, artesunate):
    - the first-line drug of choice in Thailand and used widely in many other countries

- not available in most Western countries
- highly effective, rapid acting
- resistance not reported
- minimal side-effects (neurotoxicity reported in animals)
- recrudescence   common   after monotherapy

- Multidrug-resistant infections are becoming common in parts of Thailand, Cambodia, Vietnam, Myanmar, Bangladesh, north-eastern states of India and the Amazon basin. Combination therapy should be used (i.e. artemisinin/mefloquine, atovaquone/proguanil)
- Usefulness of halofantrine is limited by its QT prolonging effect
- Intensive care is necessary in severe malaria
- Exchange transfusion efficiently reduces high parasite counts.

Prevention
- Vector control using insecticides and larvicides
- No vaccine has been convincingly shown to be protective, but trials are in progress
- Protection from biting is essential by use of knock-down insect sprays, mosquito bed-nets, insect repellents, mosquito coils, sensible clothing covering limbs, and avoiding exposure at night
- Drug prophylaxis protects against developing disease. The choice depends upon the likelihood and type of malaria, countries (and areas within) being visited, the duration of intended stay, the prevalence of chloroquine-resistant P. falciparum, and host factors (age, pregnancy, medical contraindications to certain drugs). The following are the most commonly used regimens in the major malarious areas:
  - chloroquine/paludrine (south Asia, Central America)
  - mefloquine or doxycycline or Malarone (Africa, south-east Asia, South America, Bangladesh)
- Chemoprophylaxis should start 1 week before departing and continue for 4 weeks after returning (1 week for Malarone)
- Standby therapy is appropriate for travellers with symptoms suggestive of malaria in remote

malarious regions without easy access to medical facilities, especially in south-east Asia. Fansidar, Malarone and artemisinin are the most important drugs in this category. Treatment for malaria should not be considered until at least 10 days have elapsed after entering a malarious area.

Prognosis
Mortality is exceptionally rare in P. vivax, P. ovale and P. malariae infection. For patients with complicated P. falciparum infection, mortality approaches 10–20%. In those who survive cerebral malaria, about 5% of adults and 10% of children have neurological sequelae.

South American trypanosomiasis (Chagas' disease)

Epidemiology
Chagas' disease is a zoonosis resulting from infection with Trypanosoma cruzi and is prevalent in poor, rural areas of Central and South America.

- Chronic Chagas' disease is a major health problem, affecting over 20 million persons. Up to 10% of suburban populations in Brazil are infected with the parasite
- Acute infections occur before 10 years of age in 85%; 25% will go on to develop chronic disease at 35–45 years of age
- It is geographically variable in presentation (e.g. gastrointestinal tract involvement is uncommon in Venezuela)
- It is transmitted by reduviid bugs which feed on a wide range of mammalian species:
  - bugs are infected during feeding on infected host with circulating trypomastigotes which multiply within its gut
  - transmission to human/animal hosts occurs when mucous membranes (most frequently the conjunctiva) or breached skin are contaminated by bug faeces which are produced during feeding
- It is frequently transmitted by contaminated blood during transfusion (10 000–20 000

cases/year in Brazil alone). Rarely, transmission may occur *in utero* (3% of infected mothers), during breast-feeding, from organ transplantation or nosocomially
- It may reactivate in immunosuppressed patients
- The IP is 1–4 weeks.

Pathology and pathogenesis
- Following inoculation, pseudocysts develop which are collections of intracellular amastigotes, which then rupture releasing trypomastigotes into the blood
- These are readily detectable during the acute stage
- Lymphomononuclear infiltration is detectable around clustered parasites in the myocardium and myenteric plexus, where neuronal denervation occurs
- In chronic disease, there is cardiac dilatation, apical aneurysm formation, mural thrombi and chronic lymphocytic infiltration; amastigotes are rarely found. Pathogenesis is incompletely understood. There is some suggestion that *T. cruzi*, heart muscle and nerve fibres share a common antigen which stimulates an autoimmune humoral and cellular response.

Clinical features
Acute stage
- A local lesion (chagoma, palpebral oedema) develops at the portal of entry in 50%
- Regional lymph nodes are enlarged
- Systemic manifestations are common: fever, anorexia, malaise, myalgia and headache
- On examination, there may be:
  - hepatosplenomegaly and generalized lymphadenopathy
  - oedema of the face and lower limbs
  - pronounced tachycardia, indicating myocarditis.

Chronic disease
- Occurs in 25% of patients, but echocardiography (ECG) is abnormal in 60%
- Symptoms relating to cardiac failure (dyspnoea), arrythmias (dizziness) and thromboembolic episodes predominate

- Dysphagia, reflux oesophagitis, pain and regurgitation of undigested food are the features of megaoesophagus, and constipation may suggest megacolon.

Diagnosis
Confirmation of acute Chagas' disease is by demonstration of the causal agent by:
- microscopy of anticoagulated blood or buffy coat for motile trypomastigotes; alternatively, of stained thin and thick blood smears to demonstrate parasites
- isolation of the agent by:
  - inoculating blood into mice or NNN medium
  - xenodiagnosis, which involves using laboratory-reared reduviid bugs to feed on the patient's blood. These are then examined for parasites 30 days later.

Serology is the diagnostic method of choice in chronic disease.

Treatment
Treatment is effective if given early in the acute stage and should be initiated on clinical suspicion if diagnostic delay is likely. Two agents are available and are given for 2 months:
- nifurtimox or
- benznidazole.

For chronic disease, complications should be treated appropriately. Surgery has a role in gastrointestinal tract disease. Specific treatment has no place.

Prognosis
In acute disease, mortality is highest in congenital cases, immunosuppressed patients and young children. In areas of Brazil, cardiac disease is the most common cause of death in young adults. When disease results in heart failure, half will die within 2 years.

Prevention
- Vector control by insecticide spraying
- Housing improvement, health education, etc.
- Screening donated blood (or treating with gentian violet)
- No vaccine is available.

African trypanosomiasis

Epidemiology
African trypanosomiasis is a protozoal infection restricted to localized areas within Africa and caused by two epidemiologically distinct subspecies of *Trypanosoma brucei*: *T. brucei gambiense* (west and central Africa) and *T. brucei rhodesiense* (east Africa).
• Infection is transmitted by various species of tsetse flies. Their habitats are close to water (riverine—west African) or open country (savannah—east African)
• At least 20 000 cases occur annually; it is estimated that 50 million persons are at risk
• The principal mammalian hosts are humans (west African) and animals (bushbuck—east African)
• Tsetse flies become infected during a blood meal on an infected mammalian host. Within the insect gut parasites mature and multiply, and migrate to salivary glands
• Infected tsetse flies inject trypomastigotes into mammalian skin tissues during a blood meal
• West African trypanosomiasis is likely in rural communities with river water contact
• East African trypanosomiasis is a sporadic disease of persons in contact with the animal reservoir (e.g. safari park visitors)
• Disease tends to be less severe, more chronic and associated with a lower-grade parasitaemia if due to *T. brucei gambiense*
• The IP is 2–28 days, usually shorter in east African disease.

Pathology and pathogenesis
There are two stages: haemolymphatic (stage 1) and meningoencephalitic (stage 2).
• Following inoculation, there is a proliferation of trypanosomes with a lymphocytic inflammatory infiltrate. Trypanosomes then disseminate (stage 1), leading to systemic symptoms and reticuloendothelial hyperplasia. The characteristic cell is the morular cell of Mott (foamy plasma cell)
• Stage 2 is a diffuse meningoencephalitis with trypanosomes identifiable in the brain and spinal cord, associated with a lymphocytic and morular cell infiltrate. The cyclical fever pattern is a result of variations in the surface glycoprotein and allows the parasite to evade host immune responses more effectively.

Clinical features
In the acute stages of illness
• An indurated, painful chancre at the bite site usually develops (much more frequently with east African trypanosomiasis); regional lymphadenopathy often occurs
• Fever and influenza-like symptoms are typical. The fever may develop into a pattern of remission and relapse, each phase lasting approximately 1 week
• A transient macular rash may be seen
• Hepatosplenomegaly and mobile rubbery lymphadenopathy develop, especially suboccipitally (Winterbottom's sign—more common in *T. brucei gambiense*).

In the late stages of illness
• Central nervous system (CNS) involvement is common and characterized by intractable headache, choreoathetoid movements, ataxia, behavioural abnormalities, cranial nerve lesions, deep hyperaesthesia and later drowsiness, confusion and somnolence (sleeping sickness)
• Endocrine dysfunction, weight loss and anaemia may occur.

Complications
• Myocarditis, pericardial effusion
• Pulmonary oedema
• Acute fulminant presentation (mimicking severe falciparum malaria).

Diagnosis
This is made by:
• wet preparations or thick and thin Giemsa stains of peripheral blood or buffy cells
• stained aspirates from chancres, lymph nodes, bone marrow or cerebrospinal fluid (CSF)

• serology. This is useful in non-indigenous persons and in seroprevalence surveys, but not in indigenous persons where it may reflect past exposure
• CSF examination. This is essential in all patients with trypanosomiasis, as it determines treatment. Typically in stage 2 CNS disease, there is a raised pressure, >5 lymphocytes/μL, raised total protein, raised total IgM and morular cells
• culture and animal inoculation.

Treatment
The choice of drug depends on the stage of infection and the infecting species.
• Acute stage:
  • *T. brucei gambiense* — pentamidine isethionate
  • *T. brucei rhodiense* — suramine
• Late stage — melarsoprol for either species. Corticosteroids may reduce the development of reactive encephalopathy during melarsoprol therapy.

Prevention
• Effective surveillance and treatment schedules
• Vector control with insecticides and avoiding bites
• Appropriate land use.

Prognosis
• Acute-stage disease can be cured, although fulminant infection as seen in non-indigenous travellers has a significant mortality
• Late-stage disease has a greater morbidity and mortality, with less effective and more toxic treatment. Reactive encephalopathy induced by melarsoprol occurs in 5% and has a mortality rate of 50%.

Leishmaniasis

Leishmaniasis is a vector-borne disease caused by the protozoan genus *Leishmania*, an intracellular parasite, and can be divided into cutaneous, visceral (kala-azar) and mucocutaneous syndromes.
• Transmission is through the bite of female phlebotomine sandflies which inject the infective stage, promastigotes, during blood feeding
• Promastigotes transform into amastigotes inside mammalian hosts, multiply and affect different tissues depending on the species involved
• Sandflies are infected when feeding on an infected mammalian host with amastigotes in the bloodstream. They are intracellular pathogens
• With the exception of *L. donovani* and *L. tropica* in India, the disease is a zoonosis with the major reservoirs being rodents and dogs.

Old-world cutaneous leishmaniasis
Epidemiology
Leishmania tropica, *L. major* and *L. aethiopica* are the causes of a self-limiting, indolent ulcer affecting exposed sites. This form of cutaneous leishmaniasis:
• is found in central and south-western Asia, Mediterranean, the Indian subcontinent, China and sub-Saharan Africa
• embraces three subspecies: *L. major* (rural, desert rodent reservoir, multiple ulcers); *L. tropica* (urban, human and dog reservoir, single ulcer); and *L. aethiopica* (rural, rock hyrax reservoir, non-ulcerative)
• is influenced by the host's immune response
• is a sporadic disease with occasional outbreaks during road construction, etc.
• usually has an IP of 2–4 weeks (rarely up to 3 years).

Pathology and pathogenesis
• Amastigotes within macrophages at the bite site elicit a granulomatous inflammatory reaction
• Eradication and immunity occur over 6–12 months and the patient develops cutaneous hypersensitivity
• Where immunity fails to develop, the organisms disseminate through the skin without ulceration (diffuse cutaneous leishmaniasis)
• Where the immune response is excessive,

a papular reaction over and around the scar develops (leishmania recidivans).

Clinical features
• A painless papule forms 2–4 weeks after the bite, which enlarges over 3 months into a circular ulcer with raised, indurated edges and satellite lesions
• Most heal spontaneously over 1 year, leaving a flat, depigmented atrophic scar
• Regional lymphadenitis is common.

Complications
• Diffuse cutaneous leishmaniasis
• Leishmania recidivans.

Diagnosis
• Differential diagnosis of the ulcerated form includes an infected mosquito bite, tropical ulcer, cutaneous diphtheria, *Mycobacterium ulcerans*, sporotrichosis and amoebic ulcer
• Differential diagnosis of the papular form includes lepromatous leprosy, tuberculosis and sarcoidosis
• Specimens are collected by split-skin smears, curettings or biopsy of the ulcer edges
• Leishmaniasis is confirmed by identification of the organism by Giemsa smear, or culture on NNN medium
• The leishmanial skin test is helpful in non-indigenous persons and becomes positive after 2 months.

Treatment
• No treatment is usually needed
• For cosmetically unsightly or large multiple lesions, local therapy with paromomycin cream, sodium stibogluconate injections or cryotherapy can be used
• Rarely, parenteral agents are indicated (e.g. for diffuse cutaneous leishmaniasis) (see 'Visceral leishmaniasis' below).

Prevention
• Avoidance of being bitten (sensible clothing, insect repellent, building dwellings away from forest, bed-nets)

• Insecticide sprays in houses; control of sand-fly breeding
• Control of the animal reservoirs (dog licensing and destroying strays; rodent control).

Prognosis
Old-world cutaneous leishmaniasis is a benign disease. Both diffuse cutaneous leishmaniasis and leishmania recidivans are chronic illnesses lasting 20 or more years.

Visceral leishmaniasis (kala-azar)
Epidemiology
• Visceral leishmaniasis is a chronic systemic disease caused by *L. donovani* (India and east Africa), *L. infantum* (Mediterranean) and *L. chagasi* (Central and South America)
• The reservoirs are dogs (Central and South America, Mediterranean, China, Asia), rodents (east Africa) and humans (India)
• Disease may be sporadic (non-indigenous, adults), endemic (indigenous, children <5 years) or epidemic (indigenous, all ages)
• Asymptomatic infections are common
• It is an important opportunistic infection among HIV-infected individuals with severe immunosuppression (either newly acquired or reactivation of old dormant infection). Most reported cases are in southern Europe and are due to *L. infantum*
• It affects children and young adults
• The IP is 2–4 months.

Pathology and pathogenesis
• Following inoculation, amastigotes disseminate to the reticuloendothelial system where they result in enlargement (spleen, liver, lymph node) or replacement (bone marrow)
• Polyclonal humoral stimulation occurs, resulting in hyperglobulinaemia but with defective cell-mediated response and an inability of T cells to activate macrophages to kill the parasite.

Clinical features
• In the classic form of the disease:
    • there is an insidious onset with fever (twice-daily peak), sweating and malaise

- weight loss, diarrhoea, a dry cough and epistaxis then develop
- hepatosplenomegaly, pancytopenia and wasting develop over 1–2 months; the spleen may become enormous
  - lymphadenopathy may be present
  - with progressive emaciation, the skin may darken, hair fall, and petechiae and bruising develop
- In immunocompromised HIV patients, manifestations can be atypical with cutaneous or gastrointestinal tract involvements.

Complications
- Intercurrent infection (e.g. pneumonia, tuberculosis)
- Internal haemorrhage
- Renal amyloidosis
- Post-kala-azar dermal leishmaniasis:
  - 1–3% (Africa) and 10–20% (India) of patients develop post-kala-azar dermal leishmaniasis 1–5 years after treatment, restricted to the skin (especially the face) with progressively hypopigmented patches, butterfly erythema amd multiple nodules developing
- Mucosal spread.

Diagnosis
- The differential diagnosis is wide and includes typhoid, brucellosis, lymphoma, disseminated tuberculosis, tropical splenomegaly syndrome, hepatic schistosomiasis and lepromatous leprosy
- Non-specific features supporting a diagnosis of visceral leishmaniasis are:
  - normocytic normochromic anaemia, leucopenia and thrombocytopenia
  - raised IgG, total globulin and erythrocyte sedimentation rate (ESR) with a low albumin
- Confirmation is by:
  - demonstration of amastigotes (Leishman–Donovan bodies) from lymph node (60%), bone marrow (85%) or splenic (98%) aspirate, liver biopsy (60%) or buffy coat cells
  - culture on NNN medium
  - serology (less sensitive in the immunosuppressed).

Treatment
Parenteral therapy is necessary.
- Stibogluconate (resistance may occur) or amphotericin B (liposomal form less toxic) are the standard treatments
- Pentamidine isetionate or paromomycin are alternative therapies.

Prevention
- Measures against vector
- Control of the animal reservoir
- Control of the human reservoir by prompt treatment of human cases and case-finding.

Prognosis
Prognosis is good but, without treatment, nearly all patients die within 2 years. Follow-up is imperative to detect relapses which occur in up to 25% of cases.

New-world cutaneous leishmaniasis
Epidemiology
- New-world leishmaniasis causes single or multiple ulcers (identical to old-world *L. tropica* infection) and mucocutaneous disease
- Relatively rare and occurs in Central and South America
- Two species account for the benign, self-limiting ulcers: *L. mexicana* and *L. peruviana*
- *Leishmania braziliensis* infections often result in mucocutaneous disease
- The primary reservoirs are the dog (*L. peruviana*) and rodents (*L. mexicana*, *L. braziliensis*); domestic animals may serve as secondary reservoirs
- The IP is 2–8 weeks for localized cutaneous disease and 1 month to ≥ 1 year for mucocutaneous disease.

Pathology and pathogenesis
The pathogenesis of localized cutaneous ulcers and diffuse cutaneous leishmaniasis is the same as for *L. tropica* infections. Mucocutaneous disease is associated with scarce amastigotes and prominent mononuclear cell infiltrate.

Clinical features
- Mucocutaneous disease is characterized by

the reappearance of mucosal lesions months to years after the disappearance of the primary ulcer
• The mucosal disease results in a mutilating and destructive process affecting the nose, oral cavity and pharynx.

Complications
• Aspiration pneumonia
• Airways obstruction.

Diagnosis
• The differential diagnosis is wide and includes lepromatous leprosy, South American blasto-mycosis, yaws, neoplasms, sporotrichosis and tertiary syphilis
• Non-specific features supporting a diagnosis of new-world leishmaniasis are:
    • history of residence in an endemic area
    • classical destructive features of mucocuta-neous disease
• Confirmation of new-world leishmaniasis is through:
    • identification of amastigotes
    • isolation of Leishmania on NNN medium
    • positive serology (60–95%), although there is some cross-reaction with Chagas' disease
    • a positive Leishmania skin test (85–100%).

Treatment
• Sodium stibogluconate or amphotericin B
• Relapse rate is high with in advanced disease.

Prevention
Vector control, insect repellents and control of animal reservoirs.

Prognosis
In its worst form, this is an inexorably progres-sive and mutilating condition which is occasion-ally fatal.

Toxoplasmosis

Epidemiology
• The causative agent Toxoplasma gondii, a pro-tozoan parasite, affects most mammals, includ-ing humans

• Members of the cat family are the only known definitive hosts (support the sexual cycle of T. gondii) and cats are the main reservoir for human infections.
• Oocysts excreted in cat faeces are ingested by many animals (e.g. mice and birds), where the asexual cycle takes place in tissues, which are subsequently eaten by the cat.
• Humans are infected:
    • by ingesting oocysts (common)
    • by eating undercooked meat (uncommon)
    • transplacentally
    • very rarely at transplantation (particularly cardiac), by blood transfusion or as an occu-pational hazard by laboratory workers
• Infection is highly prevalent worldwide, com-moner in warmer climates
• Seropositivity increases with age (UK preva-lence of 8% in children < 10 years rising to 47% in those > 60 years), with a seroconversion rate of 0.5–1%/year
• 10–15% of women of childbearing age in the USA are positive
• The high prevalence in France (85%) is proba-bly related to higher consumption of under-cooked meat
• Infects 0.3% of pregnant women in the UK during pregnancy, in 30% of whom the fetus is in-fected. In utero transmission is lowest in the 1st trimester (15%) and highest in the 3rd (60%); severity is greatest when infection occurs in the 1st trimester. However, congenital toxoplas-mosis is uncommon: less than 20 cases are re-ported each year in England and Wales
• Causes subclinical infection in 80–90%. It is estimated to cause 5% of cases of clinically sig-nificant lymphadenopathy
• It is an major cause of neurological disease in severely immunocompromised HIV-infected patients
• Tissue cysts persist for the life of the host and are the source of recrudescence
• IP is 5–23 days.

Pathology and pathogenesis
• Following entry, the parasites disseminate widely via blood and lymphatics; any tissue or organ may be invaded
• The infection is checked by cytokine-induced

macrophage-mediated destruction; later, humoral immunity is important in the induction of the cyst formation. *Toxoplasma* trophozoites have the ability to prevent lysosome–phagosome fusion and acidification of the lysosome within macrophages
- The encysted *T. gondii* parasites remain viable until the death of the host
- The characteristic feature on lymph node biopsy is a stellate abscess
- In congenital infection, tissue necrosis results in microcephaly, choroidoretinitis and intracranial calcification.

Clinical features

Clinical disease may present at birth (congenital), or in a young adult as lymphadenitis (acquired) or choroidoretinitis (congenital).

Congenital toxoplasmosis
- *Toxoplasma* acquired during pregnancy may lead to spontaneous abortion, stillbirth and premature delivery
- The majority of infected live offspring do not have detectable disease at birth but do carry a significant risk of developing ocular or neurological disease in later life. Those with overt clinical disease as neonates tend to progress to severe sequelae.

Toxoplasmal lymphadenitis
- This presents with:
  - lymphadenopathy, mainly cervical and usually non-suppurative and non-tender
  - an influenza-like illness with low-grade fever
  - occasional hepatosplenomegaly
- The symptoms and signs may wax and wane, but resolve within a few months.

Ocular toxoplasmosis in adults
- A third of all cases of choroidoretinitis in adults are believed to be caused by *T. gondii*
- Most are believed to be congenitally acquired
- It is a focal necrotizing retinitis which heals with clearly demarcated edges enclosing pale atrophic retina with black pigment. Symptoms include blurred vision, pain and photophobia.

Cerebral toxoplasmosis
See p. 224.

Diagnosis
- Non-specific features supporting a diagnosis of toxoplasmosis are:
  - history of contact with cats or ingestion of raw meat
  - heterophile antibody-negative glandular fever-like illness
  - typical histology on lymph node biopsy
- Confirmation is by:
  - serology. The tests used are the dye, latex agglutination and haemagglutination (measuring IgG antibody), and specific IgM and IgA antibody detection assays. Either a fourfold rise or fall between acute and convalescent sera in IgG or the presence of IgM or IgA antibody is needed to confirm recent infection
  - demonstration of *Toxoplasma* trophozoites in histological specimens
  - isolation of *T. gondii* (difficult)
  - DNA amplification (PCR)
- For diagnosis of cerebral toxoplasmosis see p. 224.

Treatment
- No therapy is required for lymphadenitis
- For the immunocompromised with cerebral or other systemic site for disease, and for ocular disease:
  - pyrimethamine, with sulphadiazine or clindamycin
- For congenital toxoplasmosis:
  - pyrimethamine and sulphadiazine, or spiramycin
- Steroids may be indicated in choroidoretinitis
- For pregnant women recently infected:
  - spiramycin (throughout pregnancy)
  - pyrimethamine and sulphadiazine (after 1st trimester).

Prevention
- Primary and secondary prophylaxis in HIV patients with CD4+ counts <200/mm^3: co-trimoxazole is the most suitable agent
- Identification and treatment of infection during pregnancy

- Adequate cooking of meat to render cysts non-infectious
- Avoidance of cat faeces for those at risk (immunosuppressed, pregnant women).

Prognosis
- Acquired toxoplasmosis in an immunocompetent patient is a benign illness
- Ocular disease is a rare cause of blindness
- Cerebral toxoplasmosis is a severe disease with a 10% fatality and 10% incidence of severe neurological complications.

Rabies

Epidemiology
- Rabies is a fatal encephalomyelitis in most warm-blooded hosts, including humans, caused by members of the *Lyssavirus* genus of the rhabdovirus family:
  - the classic rabies virus
  - European bat lyssavirus (EBL) 1 and 2
  - Australian bat virus
  - African Duvenhage virus
- All are closely related and are single-stranded RNA viruses
- Human infection is incidental to the reservoir of disease in wild and domestic animals
- The disease is endemic in large parts of Africa and Asia with between 40 000 and 50 000 deaths occurring annually
- In developed countries rabies is present mainly in wild animals; human infections are rare
- The UK, Antarctica, Japan, Scandinavia and many small islands have been disease free for many years
- Since 1990 wildlife animal rabies has also been eliminated in some western European countries through oral vaccination campaigns, and eventual elimination is likely from throughout western Europe
- Recently bat rabies has emerged as an important epidemiological concern in North America, Australia and Europe (though very rare)
- In recent years, adoption of improved postexposure treatment of humans and vaccination of dogs has decreased human rabies cases substantially in China, Thailand, Sri Lanka and Latin America
- In the UK:
  - the 20 or so cases of human rabies reported between 1920 and 2001 were all from animal bites abroad (usually dogs)
  - in 2002 a case of human rabies occurred in Scotland following a bat bite in a conservation area infected with EBL 2
  - EBL-infected bats (insectivorous) are found in mainland Europe, mostly in Denmark, the Netherlands and Germany, with around 600 reports between 1977 and 2000 but only three instances of human infections
  - it is very rare for EBLs to cross the species barrier so animal or human infections are rare
  - only two other bat infections have been found in the UK
- In the USA, although several distinct rabies virus variants exist in mammals human infections are very rare (1–2 annually)
- The animal reservoir varies according to geography:
  - dogs are the principal reservoir in most countries of Africa, Asia and Latin America
  - foxes and insectivore bats in Europe, Canada and Arctic regions
  - skunks, raccoons, foxes, coyotes and insectivore bats in the USA
  - wolves in western Asia
  - mongoose and jackals in Africa
  - bats (vampire/insectivore) in Central and South America; insectivore bats in Australia
- In animals:
  - infectivity is from a few days before they become unwell to death (usually within 2 weeks)
  - asymptomatic excretion of virus is very rare except in bats
  - the disease presents in two main forms: furious in dogs (excitation and aggression) and dumb in foxes (lethargy and uncharacteristic tameness), death occurring in 4–8 days
- Rabies in humans:
  - results from the bite or scratch of an animal

with virus-laden saliva; very rarely transmission may occur through mucous membranes, aerosol (bats in infested caves), or implantation (corneal transplants). The virus cannot penetrate intact skin
• is more likely to be acquired and have a shorter IP if: (i) the site of the bite is the face; (ii) the victim is young; (iii) the injuries are severe or multiple; (iv) the bite was not through clothing; (v) the animal was a wolf; and (vi) no pre- or postexposure prophylaxis was given
• causes an encephalitis where mortality approaches 100%; there have been three reported recoveries in persons who had received pre- or postexposure prophylaxis
• has an IP of 2–8 weeks (range 10 days–2 years).

Pathology and pathogenesis
• Initial amplification of virus occurs in muscle around the bite site, followed by attachment via the glycoprotein to nerve endings (probably acetylcholine receptors) and retrograde axonal transport along peripheral nerves (3 mm/h)
• The characteristic histological feature is the Negri body (a viral inclusion body)
• In later stages, virus spreads centrifugally by peripheral nerves to most tissues, including the salivary glands, where it is shed in the saliva.

Clinical features
• Rabies may be described as 'furious' or 'paralytic', depending on whether the brain or spinal cord, respectively, are predominantly affected. Initial features for both are:
  • discomfort, itching or paraesthesiae at the site of the healed bite wound
  • anxiety, agitation
  • fever, headache, myalgia, sore throat
• Furious rabies is characterized by:
  • hydrophobia (fear of drinking which brings on inspiratory muscle spasms), aerophobia (spasm precipitated by cold air) and excessive salivation
  • marked agitation and confusion but with lucid intervals
  • meningism, opisthotonic spasms, seizures,

and cranial nerve and upper motor neurone palsies
  • autonomic stimulation
• Paralytic rabies is characterized by:
  • an ascending asymmetrical or symmetrical paralysis.

Diagnosis
• The disease must be distinguished from hysteria, other forms of encephalitis, tetanus, bulbar poliomyelitis and Guillain–Barré syndrome. Non-specific features supporting a diagnosis of rabies include:
  • the presence of hydrophobia (50% of cases)
  • lymphocytic CSF with normal/slightly raised protein
  • history of animal exposure
  • the absence of muscle rigidity between spasms
• Confirmation is by:
  • demonstration of rabies antigen by antibody techniques (i.e. direct fluorescent antibody test) in corneal smear, skin biopsy from the back of the neck at the hairline or from brain biopsy
  • PCR on saliva
  • isolation of rabies virus from saliva by mouse inoculation or tissue cell culture (takes 7 days)
  • demonstration of a rising antibody titre in serum and CSF
  • demonstration of Negri bodies on brain biopsy
• Confirmation of rabies in the animal can be made by the demonstration of rabies antigen, Negri bodies on brain biopsy, or culture of rabies virus.

Treatment
• No effective treatment exists; patients should be given adequate opiate analgesia to relieve terror and pain
• Intensive care only prolongs life
• The patient should be strictly isolated and staff should wear goggles, masks and gloves in addition to being immunized (human-to-human transmission has not been documented).

Prevention

- Import controls and quarantine of imported animals
- Control of dog rabies—annual vaccination, elimination of stray dogs
- Immunization of wild animal reservoir using baits and attenuated vaccine
- Pre-exposure immunization of high-risk groups (veterinarians, wildlife conservation personnel, etc.)
- Avoidance of handling bats by unimmunized persons
- Early treatment of wounds with soap and water and virucidal antiseptic (e.g. povidone iodine, alcohol)
- Postexposure treatment (given early almost 100% effective):
  - Active immunization with a potent cell culture (e.g. human diploid cell vaccine) or an embryonated egg vaccine, given
    - intramuscularly (1 mL) on days 0, 3, 7, 14 and 28 or
    - intradermally (0.1 mL) on two sites (days 0, 3, 7) and one site (days 21 and 90)—cheaper World Health Organization (WHO)-approved alternative
  - Passive immunization with hyperimmune rabies immunoglobulin (RIG)—this can be withheld if exposure is minor (licks on skin, scratches or abrasions, minor bites through clothes on extremities), the animal is normal and under observation for 10 days, but given if the animal becomes ill and rabies is confirmed. Half of the total dose is given intramuscularly and half infiltrated into the wound
  - Treatment is stopped if animal remains healthy 10 days post bite.

Prognosis

Clinical rabies is almost always fatal. Without intensive care, one-third of victims die early during a hydrophobic spasm. The remainder lapse into coma and develop generalized flaccid paralysis.

Viral haemorrhagic fevers

Viral haemorrhagic fevers (VHFs) are a group of distinct acute viral infections that cause various degrees of haemorrhage, shock and sometimes death. Many are rarely identified and occur in geographically remote areas in the developing countries. This section deals with those VHFs that are capable of transmission from person to person and are of public health concern because of their high fatality rate, and the risk of importation into and spreading secondarily within another country through travel, i.e. Lassa, Ebola, Crimean-Congo and Marburg fevers. Dengue, yellow fever and hantaviruses can also cause haemorrhagic manifestations but do not transmit from person to person and are described separately.

Epidemiology

Lassa fever
- Caused by a single-stranded RNA virus, an arenavirus similar to LCM, and Argentinian and Bolivian haemorrhagic fever viruses, but these do not transmit from person to person
- Endemic in Guinea, Sierra Leone, Liberia and parts of Nigeria. Up to 50% of the indigenous population have serological evidence of past infection
- Results in 300 000 cases and 5000 deaths annually
- Is spread mainly by direct or indirect contact with excreta of chronically infected multimammate rats
- Person-to-person and laboratory infections occur, particularly in the hospital setting, through contact with an infected person's body fluids (blood, pharyngeal secretion, urine, semen)
- Virus may be excreted in urine and semen for many weeks after recovery
- The IP is 6–21 days.

Ebola fever
- Caused by a single-stranded RNA filovirus, first isolated in large epidemics in Sudan and Zaire (now Congo) in 1976

- In 1989, Ebola virus was identified in the USA in laboratory monkeys imported from the Philippines and caused several human infections
- It has caused epidemics and outbreaks since then in Zaire (1995), Côte d'Ivoire (1994–1995), Gabon (1994, 1996) and Uganda (2000)
- One of the most virulent viral disease of humans (fatality 50–90%)
- Transmitted by direct contact with patients' body fluids (semen may transmit for up to 7 weeks after recovery) and through handling ill or dead chimpanzees
- Appears to have a natural reservoir in the rain forests of Africa and Asia but its nature is unknown. Non-human primates are not thought to be the reservoir but like humans are infected directly or indirectly from the natural reservoir
- The IP is 2–21 days.

Crimean-Congo haemorrhagic fever (CCHF)
- Caused by a member of the *Nairovirus* group, CCHF virus primarily causes zoonoses and affects many wild and domestic animals who are infected through the bite of infected ticks (commonly *Hyalomma* ticks)
- Causes sporadic cases and outbreaks affecting humans
- Endemic in the Middle East, Africa, Balkans and central-south Asia. Recent outbreaks have occurred in Kosovo, Albania, Iran and Pakistan
- Transmitted to humans through direct contact with infected animal tissues or via tick bites, and less commonly through contact with body fluids of infected patients
- The IP is 3–12 days.

Marburg fever
- Caused by a unique filovirus, distinct antigenically from Ebola virus
- Was first identified in 1967 in German laboratory workers exposed to African green monkeys imported from Uganda
- Has been reported only in isolated cases since then in sub-Saharan Africa
- Has a natural reservoir in African rain forests but nature is unknown

- Transmitted by contact with infected animal tissues or body fluids of infected persons (including semen, which may remain infectious after recovery)
- Has an IP of 3–16 days.

Pathology and pathogenesis
Initial constitutional symptoms result from viraemia. Damage to vascular system leads to leaky capillaries leading to cutaneous and haemorrhagic manifestations. Direct organ damage may also occur, possibly through cytokine mediators.

Clinical features
- Initial presentation is often with abrupt onset of fever, myalgia and sore throat
- Soon patient develops increasing prostration with severe headache, dizziness, photophobia, suffused conjunctivae, nausea and vomiting, and diarrhoea (commoner in Ebola fever)
- Clinically patient looks acutely ill
- Petechial rash may develop
- Proteinuria is common
- Mental changes
- Lymphadenopathy (in CCHF).

Complications
- Shock and leaky capillaries (causing adult respiratory distress syndrome, pleural effusions, oedema of the face and neck, and internal and external bleeding)
- Encephalopathy
- Pericarditis and congestive cardiac failure
- Jaundice (common in CCHF).

Diagnosis
- In non-endemic countries, transmissible VHF should be suspected when a patient has:
    - history of travel to an endemic area within the last 21 days
    - been potentially exposed to reservoir (rats, sick non-human primates or working as a health-care worker)
    - influenza-like symptoms with pharyngitis, conjunctivitis and bleeding
    - negative malarial films
    - lymphopenia followed by a neutrophilia

• A restricted list of laboratory specimens for diagnosis and exclusion of treatable infections (i.e. typhoid, malaria, septicaemia) and for baseline biochemistry should be performed in a special high containment laboratory facility
• Confirmation of different VHFs is by specialized laboratory tests on blood specimens to detect specific antigens/viral genes, to isolate virus in cell culture, or to detect IgM and IgG antibodies under maximum biological containment conditions.

Treatment
• Strict isolation. If suspicion is high, the patient should be cared for in a high-security isolation unit with full-length gowns, gloves, masks, overshoes and caps
• In endemic areas suspected cases should be isolated from other patients and strict barrier nursing procedures should be observed
• IV ribavirin reduces mortality in Lassa fever if given before the 7th day (from 61% to 5%) in severe cases and possibly in CCHF.

Prevention
• Surveillance of close contacts
• Ribavirin for close contacts (of Lassa fever)
• Rodent control for Lassa fever, and tick control and personal protection against tick bites (insect repellent, protective clothing) for CCHF.

Prognosis
• Lassa fever: the overall mortality is 2%. For hospitalized cases this increases to 10–15%. There is a higher mortality in pregnancy (30%). High transaminases are predictive of a poor outcome
• Ebola fever: 50–90% mortality in clinically ill cases
• CCHF: mortality approximately 30%.

Yellow fever
Epidemiology
• Yellow fever is due to a flavivirus and is mosquito borne (principally *Aedes aegypti*)
• It is a zoonosis of monkeys in Africa and South America

• It is endemic in large areas of tropical Africa and the Americas with an estimated 200 000 cases (with 30 000 deaths) each year
• It has never been reported from Asia but the region is at risk of introduction of the virus because of the presence of appropriate monkeys and mosquitoes
• Transmission to humans takes three forms:
  • 'jungle' (sylvatic) — sporadic cases occur when humans are clearing forests with mosquitoes breeding in tree-holes. South American cases are almost exclusively 'jungle' type
  • 'urban' — large epidemics occurring when virus is introduced in densely populated areas with mosquitoes breeding in man-made containers. Occurs in Africa
  • 'intermediate' — small-scale epidemics in African savannahs during rainy season affecting many separate villages simultaneously when semidomestic mosquitoes infect humans and monkeys simultaneously: most common type now in Africa
• The IP is 3–6 days.

Pathology and pathogenesis
Following inoculation, the virus replicates in draining lymph nodes and then causes viraemia. A characteristic midzonal hepatic necrosis occurs. The kidneys show evidence of acute tubular necrosis.

Clinical features
• The illness ranges from mild to catastrophic (may be asymptomatic)
• Typically there is an acute fever with abrupt onset of severe influenza-like illness, conjunctivitis and relative bradycardia:
  • symptoms remit in 3–4 days and most remain well thereafter
  • in about 15%, toxicity returns rapidly with reappearance of fever and severe vomiting, abdominal pain, jaundice, renal failure, hypotension and haemorrhage.

Diagnosis
Yellow fever has to be distinguished from falciparum malaria, leptospirosis, fulminant viral he-

patitis, dengue haemorrhagic fever (DHF) and other VHFs. In favour of a diagnosis of yellow fever are:
- the absence of a history of immunization against yellow fever
- a history of visiting an endemic area
- a biphasic illness with hepatitis, DIC and renal failure
- thrombocytopenia and leucopenia.

Confirmation is from:
- viral antigen detection by enzyme-linked immunoadsorbent assay (ELISA)
- viral culture
- detection of IgM antibodies
- fourfold rise in IgG antibodies between acute and convalescent sera.

Treatment
Supportive treatment in intensive care is necessary for severe cases.

Prevention
- Excellent immunity can be achieved with a live attenuated vaccine (Yellow Fever 17D) which confers a long-lasting immunity (10 years at least). It is safe, but several cases of viscerotropic (multiorgan system failure) and of neurotropic disease (encephalomyelitis) with deaths, possibly a result of abnormal host response to 17D vaccine, have been reported recently in the USA, Brazil and Australia
- Vaccination is recommended for persons aged >9 months travelling to areas reporting ongoing yellow fever activity (a vaccination certification is required for entry by most countries). Risk to life from yellow fever is far greater than the risk from vaccine. Immunocompromised persons and pregnant women should not be vaccinated and should be given 'medical waiver' travel certificates
- WHO recommends routine childhood vaccination in endemic countries, and 18 African countries have agreed to adopt this. Gambia achieved 85% coverage by 2000 and has had no case since 1980 despite viral presence
- Mosquito control and measures against mosquito bites.

Prognosis
When jaundice is present, 20–60% of patients die. Death may be a result of fulminant hepatitis, renal failure, myocarditis or DIC.

Dengue fever
Epidemiology
- Dengue is the most important of the arboviruses, with 40–80 million people becoming infected worldwide each year
- 500 000 cases are hospitalized with haemorrhagic complications (dengue haemorrhagic fever, DHF)
- Prevalence has increased substantially in recent decades
- It is endemic throughout tropical Africa, the Americas, the eastern Mediterranean, India, south-east Asia and the western Pacific; the latter two areas are the most seriously affected
- Explosive outbreaks are occurring as the disease spreads to new areas with high attack rates among susceptibles
- It is a common travel-related infection in travellers from non-endemic countries
- It is transmitted by A. aegypti mosquitoes which have a predominantly urban habitat and acquire virus during blood feeding on an infected human (infective after 8–10 days)
- Humans are the main amplifying host but monkeys have been infected in some countries
- The virus is an RNA flavivirus with four serotypes, each of which can infect humans. In a country where multiple serotypes circulate, the first two infections may cause illness, but the third and fourth tend to be asymptomatic. In children (and less commonly adults), the second infection may cause haemorrhage (DHF) and shock (dengue shock syndrome (DSS))
- The IP is 3–8 days.

Pathology and pathogenesis
DHF/DSS appears to occur as a result of non-neutralizing antibody from previous exposure to a different serotype, enhancing uptake of immune complexes into monocytes. It is seen in infants who acquire dengue for the first time but who have circulating maternal antibody, and in children during a second dengue infection. DSS

is associated with cytokine release (especially TNF). DHF is associated with plasma leakage and haemorrhage.

Clinical features

Four syndromes may result from dengue virus infection: simple fever, dengue fever syndrome, DHF and DSS. The presentation is largely age dependent.
• Infants and young children infected for the first time tend to develop a simple fever
• Older children and adults infected for the first time develop dengue fever syndrome characterized by:
    • sudden onset of severe influenza-like symptoms with arthralgia (break-bone fever)
    • a maculopapular rash on the 3rd day (in some) with petechiae developing later
    • severe retro-orbital pain on eye movement or pressure
    • occasional haemorrhage
    • prolonged convalescence
• Children previously infected with primary dengue and infected with another serotype may develop DHF/DSS. In addition to the above, patients develop:
    • serous effusions, especially pleural and peritoneal
    • a haemorrhagic diathesis
    • hepatomegaly   and   generalized lymphadenopathy
    • hypotension and shock.

Diagnosis

Non-specific features supporting a diagnosis of dengue fever are:
• leucopenia with relative lymphocytosis
• negative malarial films
• thrombocytopenia and clotting abnormalities
• hypoalbuminaemia.
Confirmation of dengue infection is by:
• virus isolation from the blood
• a fourfold rise or fall in IgG antibodies
• presence of IgM antibody.

Treatment

• For dengue fever: bed rest, sponging for fever, paracetamol (avoid aspirin because of the risk of haemorrhage)

• For DHF/DSS: intensive support is necessary.

Prevention

• Avoid mosquito bites by wearing appropriate clothing and using insect repellents and mosquito nets
• Vector control: eliminating domestic breeding places and use of larvicides and insecticides
• There is no vaccine.

Prognosis

Uncomplicated dengue fever is a benign illness. The case fatality of DHF and DSS is 2%.

Diseases associated with hantavirus infections

Epidemiology

Hantaviruses are rodent viruses of the bunyavirus family (single-stranded RNA viruses) that cause a range of clinical syndromes in humans.
• Haemorrhagic fever and renal syndrome (HFRS):
    • is caused by Hantaan and Dobrava viruses
    • occurs widely in eastern Europe and far-eastern Asia with thousands of cases of severe HFRS annually
    • has a milder form, caused by Seoul virus, occurring largely in Asia
    • has reservoir in murine field mice
• Nephropathica epidemica (NE—a milder form of HFRS):
    • is caused by Puumala virus, carried by the bank vole
    • is very common in northern Europe, especially Scandinavia
• Hantavirus pulmonary syndrome (HPS):
    • is a new disease identified in 1993, occurring in North and South Americas
    • is caused by several strains of hantaviruses (sin Nombre virus—commonest) carried by sigmodontine rodents (commonly deer mouse in the USA) which are restricted to the Americas
    • several hundred cases have been reported so far in the USA
• Hantaviruses do not cause overt disease in

rodents which excrete virus in saliva, urine and faeces for months, sometimes for life
• Humans are infected through inhalation of aerosolized saliva or excreta, through direct inoculation via broken skin or conjunctiva or possibly through food or drink, or by rodent bites
• Most hantavirus infections are subclinical. Seroprevalence surveys show 10% of persons in endemic areas of Europe/Asia have antibodies
• The IP is 2–3 weeks.

Pathogenesis
There is vascular damage with capillary leak resulting in interstitial nephritis in HFRS/NE and interstitial pulmonary oedema in HPS.

Clinical features
HFRS/NE
• The clinical presentations range from a mild illness with minimal renal dysfunction, to one with severe renal failure and shock
• Classically, HFRS is divided into five clinical stages: febrile, hypotensive (day 5), oliguric (day 9), polyuric and convalescent (by day 14). Most patients present with:
  • the abrupt onset of influenza-like symptoms
  • blurred vision, conjunctivitis, flushed face, periorbital oedema and an erythematous rash
  • palatal and axillary petechia
  • back pain and tenderness from massive retroperitoneal oedema
• Later, hypotensive stage sets in with shock and haemorrhages, progressing to oliguria, laboratory evidence of DIC and renal failure

HPS
• A prodrome of fever, myalgia and nausea and vomiting, followed by
• Breathlessness and dizziness from hypotension
• Progression to severe hypoxia and respiratory failure.

Diagnosis
Suspicion should be aroused if there is:

• a history of exposure to rodents
• a combination of fever, thrombocytopenia, and acute renal failure or non-cardiogenic pulmonary oedema
• leucocytosis ($>20.0 \times 10^9$/L).
Confirmation is by:
• detection of IgM or a fourfold rise in IgG antibodies to a panel of hantaviruses
• detection of specific nucleic acid by amplification (PCR).

Treatment
• Ribavirin may be effective
• Intensive care management with renal or ventilatory support.

Prevention
• Appropriate laboratory precautions should be observed when processing patient samples
• Reducing the risk of exposure to rodents
• No vaccine is available.

Prognosis
The fatality rate is approximately 5–10% for HFRS, < 1% for NE and 37% for HPS.

Brucellosis

Epidemiology
• Brucellosis is a zoonosis of considerable public health and economic significance for the developing countries
• Of the six known species, *Brucella melitensis* (mainly affects goats), *B. suis* (pigs) and *B. abortus* (cattle) account for most human infections. *B. canis* is a problem of dogs (laboratory, kennel or stray)
• *B. melitensis* is the most pathogenic and invasive
• Human brucellosis:
  • accounts for 500 000 cases worldwide annually, with 20–30 cases in the UK every year
  • is most commonly caused by *B. melitensis* (mainly in the Middle East, Mediterranean, Latin America and Asia), followed by *B. abortus* (global)
  • in the West where animal disease has been controlled is mainly imported in patients with

a history of ingestion of unpasteurized dairy produce or rarely acquired occupationally (e.g. farm workers, laboratory staff)
• is transmitted by ingestion of contaminated meat, milk or cheese, inhalation of infected material when handling animals (at parturition or slaughter), or inoculation through broken skin/mucous membrane contamination. Very rarely, infection may be acquired sexually, vertically or via blood transfusion
• is most common in young adults, and most severe when caused by B. melitensis
• may be subclinical, as identified by serological surveys
• has an IP from 1–3 weeks to several months.

Pathology and pathogenesis
• Following ingestion or inhalation, the organisms reach the bloodstream via the lymphatics, causing bacteraemia
• They are then taken up by macrophages, in which they multiply, and localize in the reticuloendothelial system
• Once localized, they elicit a granulomatous reaction similar to that of tuberculosis, which may progress with tissue necrosis to abscess formation.

Clinical features
Onset may range from 1 day to 4 weeks. Typically, a patient with an acute or subacute illness may present with:
• influenza-like symptoms with drenching sweats, high hectic fever, rigors, myalgia and malaise. Recurring fever may develop (undulant fever)
• arthritis or arthralgia, usually monoarticular and large joint, and/or low back pain with sciatica
• headache, irritability, insomnia and confusion
• hepatosplenomegaly and lymphadenopathy.

Complications
• Arthritis, sacroileitis, vertebral osteomyelitis in 10–30% of cases
• A toxic course with haemorrhagic features

• Meningitis, encephalitis and peripheral neuritis
• Endocarditis, myocarditis, pericarditis
• Granulomatous hepatitis
• Epididymo-orchitis
• Bronchopneumonia, pleurisy.
A chronic form of brucellosis has been described, with low-grade fever, fatigue, depression, insomnia and vague rheumatic symptoms. This is usually chronic fatigue syndrome in a person who has at some time acquired asymptomatic brucellosis and has positive antibodies.

Diagnosis
Non-specific features supporting a diagnosis of brucellosis include:
• a history of consumption of raw milk or unpasteurized cheeses
• leucopenia with relative lymphocytosis
• raised ESR
• mildly deranged liver function tests (LFTs).
Confirmation of brucellosis is through:
• culture of the organism from blood, bone marrow, urine or biopsy specimens. The cultures need prolonged incubation (up to 6 weeks). A positive blood culture is found in 30–50% of B. melitensis cases, but less frequently with B. abortus
• serological tests. The standard tests are the serum agglutination test, complement fixation test, antihuman globulin test, and either ELISA or radioimmunoassay (IgM and IgG). The presence of IgM antibody, a fourfold rise or fall between acute and convalescent sera, or a single high IgG titre, indicate active infection.

Treatment
• Combination therapy for 6 weeks is necessary, with a tetracycline and an aminoglycoside being the preferred drugs
• Doxycycline is the preferred tetracycline in combination with either gentamicin or streptomycin (especially for inpatients), co-trimoxazole, ciprofloxacin or rifampicin
• Defervescence occurs within 4–5 days. A transient Herxheimer-like reaction may occur.

Prevention

- Eradication of disease in the animal population has been achieved in the UK and many other countries through test-and-slaughter and/or vaccination strategies
- Avoidance of consumption of and proper heat treatment of raw milk and milk products
- Education of those at risk, and the wearing of protective clothing, etc.

Prognosis

Most human infections are mild or subclinical, and are self-limiting over 2–3 weeks.

Bartonella infections

Bartonella species are tiny Gram-negative bacilli, several strains of which have been associated with a wide spectrum of clinical illnesses in recent years, ranging from the mild lymphadenopathy of cat-scratch disease and invasive bacteraemic illnesses of trench fever, oraya fever and endocarditis to the chronic vascular lesions of bacillary angiomatosis.

Epidemiology

- Cat-scratch disease:
  - is caused by *B. henselae* and occurs worldwide, affects children more than adults
  - human infection results from the scratch or bite of an infected cat, the asymptomatic carriers of infection, probably transmitted directly from other cats or through flea bites
- Trench fever is caused by *B. quintana* or *B. henselae*, occurred in epidemics during world wars among soldiers, and occurs sporadically worldwide in conditions encouraging human body louse, the transmitter of infection from person to person
- Oraya fever is caused by *B. bacilliformis*, is transmitted by sandflies and is endemic in the river valleys of the Andes mountains
- Bacillary angiomatosis:
  - is caused by *B. henselae* and *B. quintana* and occurs primarily in immunocompromised persons, particularly those infected with HIV

- cats are the probable reservoir, cross-infected via fleas
- Endocarditis has recently been associated with *B. elizabethae* and *B. henselae*.

Pathology

Varies from localized granulomatous inflammation in cat-scratch disease, to new blood vessel formation and neutrophilic inflammatory response in bacillary angiomatosis, erythrocyte invasion in oraya fever and bacteraemia in trench fever.

Clinical features

- Cat scratch disease:
  - papular lesion, later pustulating at the site of cat scratch or bite and associated regional lymphadenopathy (may suppurate)
  - malaise and anorexia
  - lymph node enlargement persisting for weeks or months without treatment
- Trench fever — high, often paroxysmal, persistent fever, meningism
- Oraya fever — fever, severe haemolytic anaemia, headache and body aches
- Bacillary angiomatosis:
  - red or purple nodular swellings of varying sizes; may resemble Kaposi's sarcoma occurring on skin or mucous membrane
  - disseminated lesions may occur in liver (peliosis hepatis), spleen, lymph nodes, bone marrow or CNS
  - persisting fever and weight loss common with disseminated lesions.

Diagnosis

- Typical histology and demonstration of bacilli with Warthin–Starry silver stain are diagnostic in cat-scratch disease and bacillary angiomatosis
- CT or MR scans help localizing deep nodular lesions
- Serology is useful in suspected cat-scratch disease but not in the immunocompromised
- Positive blood culture in trench and oraya fevers
- PCR is promising and may help when other tests are unrevealing.

Treatment
• Bartonellae are susceptible to a wide range of antimicrobials including penicillin, tetracycline, chloramphenicol and macrolides
• Course should be prolonged (3–4 weeks).

Rickettsial diseases

Epidemiology
• *Rickettsia* have mammalian reservoirs (humans included), are transmitted by arthropods (ticks, mites, lice or fleas), and are distributed in geographically distinct areas
• The major *Rickettsia* are *R. prowazeki* (louse-borne (epidemic) typhus), *R. typhi* (murine (endemic) typhus), *R. tsutsugamushi* (scrub typhus), *R. rickettsii* (Rocky Mountain spotted fever (RMSF)), *R. conori* (tick typhus) and *R. akari* (rickettsial pox).

Louse-borne typhus
• May cause epidemics in poor socioeconomic conditions
• A disease of communities that do not change their clothes (e.g. cold, poverty, war, famine; now a disease of tropical highlands); it is mostly reported from Africa
• Transmitted by the body louse (*Pediculus humanus*) when louse faeces infected with *R. prowazeki* contaminate the bite wound or mucous membranes
• Has a reservoir of infection in humans (epidemics) and flying squirrels (sporadic cases)
• May recrudesce as Brill–Zinsser disease.

Murine typhus
• An urban zoonosis with a reservoir in rats
• Transmitted by the tropical rat flea (*Xenopsylla cheopis*) when flea faeces infected with *R. typhi* contaminate the bite wound or mucous membrane
• Endemic in southern USA, the Mediterranean, the Middle East, the Indian subcontinent and south-east Asia.

Scrub typhus
• Transmitted by the bite of the larval stage of trombiculid mites
• Widely distributed in the Asiatic-Pacific region
• A rural disease
• Transferred transovarially in the mites
• Commonly infects small mammals.

Spotted fevers (RMSF, tick typhus)
• Can be caused by a range of *Rickettsia*, depending on the location: *R. rickettsii* (North and South America) to *R. conori* (Africa, Mediterranean)
• Transmitted by ticks whose bites often go unnoticed.

The IP for the rickettsial infections varies between 1 and 2 weeks, but occasionally may be as short as 2 days or as long as 18 days.

Pathology and pathogenesis
• The prime pathological abnormality is a widespread vasculitis caused by rickettsial invasion of endothelial cells
• All organs may be involved, but particularly the skin, brain, kidneys and heart
• Vasculitis causes impaired vascular integrity and increased permeability giving rise to oedema, hypovolaemia, hypotension and hypoalbuminaemia
• Perivascular inflammation, endothelial cell proliferation, arterial wall necrosis and thrombotic occlusion of arterioles occur.

Clinical features
The illness may pass unnoticed or be ascribed to a viral infection. The typical features of rickettsial disease are:
• abrupt onset of malaise, headache, fever, rigors and vomiting
• an eschcar at the site of the biting vector (scrub and tick typhus (50%) and rickettsial pox (95%)). It is a painless, punched-out ulcer covered with a black scab. There may be associated tender lymphadenopathy
• a maculopapular or petechial rash. Characteristic features of this rash are as follows:

- it is most prominent on the extremities in RMSF and tick typhus where it affects the palms and soles, and on the trunk in scrub, endemic and epidemic typhus
- it develops from being maculopapular into a petechial/purpuric rash (except in scrub typhus); in severe cases it is accompanied by ischaemia of the extremities
- it is vesicular in rickettsial pox
- it usually develops on days 3–5.

Complications
- Aseptic meningitis, encephalitis, optic neuritis, deafness
- Renal failure
- Non-cardiogenic pulmonary oedema, myocarditis, pneumonitis
- Recrudescence (*R. prowazeki*).

Diagnosis
- RMSF and tick typhus have to be distinguished from measles, meningococcal septicaemia, vasculitis and idiopathic thrombocytopenic purpura, and early typhus from typhoid, malaria and relapsing fever
- Non-specific features supporting a diagnosis of rickettsial infection are:
  - potential exposure to a vector
  - a characteristic rash
  - normal white cell count (WCCt) and ESR
- Confirmation is by:
  - direct immunofluorescence of a skin biopsy (RMSF)
  - serology on acute and convalescent sera showing a fourfold rise in IgG antibodies or detection of IgM antibody
  - isolation of *Rickettsia* from blood. This is hazardous, so is rarely attempted
  - nucleic acid amplification (PCR).

Treatment
- *Rickettsia* are insensitive to cell wall antibiotics such as penicillins and cephalosporins. Active antibiotics include tetracycline (e.g. doxycycline), chloramphenicol or ciprofloxacin
- The patient usually improves rapidly with appropriate therapy, with defervescence within 2 days
- Intensive care is necessary for patients with severe vasculitis.

Prevention
- Vector (e.g. delousing with DDT) and reservoir (e.g. rat-curbing measures) control
- Personal measures to reduce insect bites, such as appropriate clothing and insect repellents
- Chemoprophylaxis with weekly doxycycline works in protecting against scrub typhus infection
- There are no vaccines available.

Prognosis
Epidemic louse-borne typhus and RMSF are the most serious. If antibiotic therapy is started before complications have occurred, fatalities are rare.

Leptospirosis

Epidemiology
- Leptospirosis is a zoonosis caused by a spirochaete with worldwide distribution
- It is commoner in temperate or tropical climates
- The genus *Leptospira* has been divided into two species: pathogenic (*L. interrogans*) or free-living (*L. biflexa*)
- 25–50 cases occur in the UK annually
- 202 pathogenic serovars of *L. interrogans* exist worldwide; 16 have been isolated in the UK
- Severity of illness is related to serovar, but not absolutely (e.g. Weil's disease is usually caused by *L. icterohaemorrhagiae*)
- A wide range of animals (e.g. cattle, pigs, horses, dogs, rodents and wild animals) form the natural reservoir; they may become ill or remain asymptomatic
- Worldwide human infections are caused most frequently by *L. icterohaemorrhagiae, L. pomona, L. canicola, L. autumnalis, L. grippotyphosa, L. balam, L. hardjo* and *L. australis*

- The predominant UK strains are *L. icterohaemorrhagiae* (rat) and *L. hardjo* (cow)
- Pathogenic leptospires can live for 4 weeks in fresh water, 6 months in urine-saturated soil and 24 h in seawater
- Transmission is when damaged skin or mucous membranes come into contact with infected animal urine
- Exposure usually occurs as a result of occupation (e.g. farmers, sewer workers, veterinarians, fishery workers) or recreational pursuits (e.g. water sports)
- Most cases occur in working men and are reported during summer and autumn
- Serological surveys of at-risk populations show it to be an infrequent infection
- The IP is 10 days (range 5–19 days).

Pathology and pathogenesis
- Leptospiraemia is followed by localization in many areas, including the CSF
- Vasculitis with capillary injury and haemorrhage results from immune complex deposition
- Jaundice is a result of hepatocellular dysfunction; biopsy shows no evidence of hepatocyte necrosis
- Renal failure is primarily a result of tubular damage with acute tubular necrosis
- During the leptospiraemic phase, CSF cultures are positive for leptospires but without any inflammatory reaction; this follows in the second phase when the cultures become negative.

Clinical features
The illness follows a biphasic pattern, the first phase representing the leptospiraemia and lasting 4–5 days. It is characterized by:
- abrupt onset of high fever, rigors and headache
- intense myalgia and muscle tenderness
- occasional confusion, abdominal pain, nausea and vomiting, cough, chest pains, maculopapular rash and haemoptysis.

In many patients, the disease may not progress to a second phase and the patient recovers, having had an influenza-like illness. Features of the second phase include:

- recurrence of fever after 2–3 days
- suffused conjunctivae
- aseptic meningitis (Canicola fever) with a lymphocytic CSF, normal/raised protein and normal glucose
- deep jaundice, renal compromise and proteinuria (Weil's disease)
- pretibial raised erythematous rash (Fort Bragg fever).

Complications
- Renal failure—anuria is common in Weil's disease but short-lived
- Myocarditis—usually presents early with arrhythmias
- ARDS and DIC with fulminant progression to a multisystem disease; most deaths occur around 14 days
- Chronic uveitis
- Relapse.

Diagnosis
The initial diagnosis is clinical, with the triad of high fever, renal failure and deep jaundice having few other causes. Non-specific features supporting a diagnosis of leptospirosis include:
- the abrupt onset, severe myalgia and suffused conjunctivae
- a leucocytosis with a neutrophilia on differential WCCt
- a lymphocytic CSF
- a raised creatine phosphokinase (5×normal in 50% of cases)
- relatively normal LFTs. Typically there is a very high bilirubin level, with the transminases and alkaline phosphatase levels at between 1 and 5×normal (compared to 100×normal in viral hepatitis)
- raised urea and creatinine, together with proteinuria.

Confirmation of leptospirosis can be demonstrated by:
- detection of specific antibody. This becomes positive by 7 days, the usual time of clinical presentation
- culture (takes weeks and is insensitive)
- identification of leptospires by dark-ground

illumination (technically difficult and false-positive rates are high)

• PCR which can detect infection in serum, urine and other body fluids and identify serovars.

Treatment

• All recognized cases should be treated. Mild infections recover without specific treatment and may only be recognized retrospectively

• For severe or complicated disease, IV benzylpenicillin is the treatment of choice. It is effective up to the 2nd week of the illness. Tetracycline or doxycycline are alternatives

• Dialysis is usually needed in patients developing renal failure.

Prevention

• Education of those at risk in reducing likelihood of infection

• Immunization of domestic livestock and pets

• Prophylaxis with doxycycline if exposure is likely

• Control of the reservoir (e.g. rats)

• No human vaccine is available.

Prognosis

Death is very rare and is usually due to massive haemorrhage, acute renal failure or, occasionally, cardiac failure.

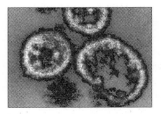

Helminthiasis

Helminths are parasitic worms, which infect all forms of animal and plant life. Many are pathogenic to humans; only the common or important ones are described here.

Nematode infections

These are roundworms. Human infections take two forms.
• Intestinal nematode infections—life cycle only involves humans; adult worms live in their intestines
• Tissue nematode infections—adult worms are animal parasites and live in their intestines; humans develop tissue infections from larval invasion if involved in life cycle.

Intestinal nematode infections
Enterobiasis (threadworm or pinworm infection)
Epidemiology and life cycle
• The parasite *Enterobius vermicularis* affects humans only; it is the most common worm infection worldwide
• Usually affects children. Often several members of a household or institution are infected
• 1 cm long adults live and mate in the colon and survive only for 6 weeks
• Gravid females lay thousands of eggs at night in the perianal area

• The eggs are then carried by hands, to be ingested by the same person or by new hosts
• Eggs change to larvae and mature into adults in the intestine, completing the cycle in about a month.

Clinical features and diagnosis
• Perianal itching during egg deposition is the only symptom
• White, slightly motile, gravid female thread-like worms may be visible in the faeces
• 'Persisting' or relapsing infections are always due to reinfection, as adult worms can live only for 6 weeks and cannot multiply within the intestine
• Diagnosis is by microscopic demonstration of trapped eggs on a sticky cellophane tape applied to the perianal skin in early morning and then placed on a glass slide or by visualizing the worm.

Treatment
• Mebendazole, as a single dose, is the treatment of choice in anyone above 2 years
• In children <2 years piperazine for 7 days is preferred
• The whole family and close sexual contacts require treatment
• Daily morning baths and hand-washing after defecation

- A second course of treatment after 2 weeks is advisable as reinfections are common.

Ascariasis (common roundworm infection)

Epidemiology and life cycle

- Humans are the only host for *Ascaris lumbricoides*
- Not endemic in developed countries but can be found in recent immigrants from developing countries where it is very common. Most cases occur in children
- Adults live in the small intestinal lumen, survive for a year, and can reach up to 40 cm in length
- Gravid females lay thousands of eggs a day which pass on to soil with faeces, become infective in few weeks and can survive for long periods in the soil
- Transmission is by ingestion of faecally contaminated soil, via hand or vegetable produce
- Larvae hatched in jejunum penetrate into mucosal circulation and migrate via the liver, lungs, bronchus and trachea to the pharynx and are swallowed into the intestine, where they mature into large, pale pink, round worms
- Multiplication does not take place within the intestine.

Clinical features and diagnosis

- Most infected individuals have low worm burdens and are asymptomatic

- In heavy infections intestinal obstruction may develop
- Worms entering the common bile duct cause biliary obstruction
- Visceral larva migration may cause tender hepatomegaly or allergic pneumonitis and an intense eosinophilia
- Passage of adult worms or the finding of ova in faecal smears or concentrate are diagnostic
- Barium meal may reveal the worm (Fig. 18.1).

Treatment

- Levamisole is very effective, well tolerated and the drug of choice and is currently available in the UK on a named-patient basis
- Mebendazole and piperazine are also effective.

Ancylostomiasis and necatoriasis (human hookworm infection)

Epidemiology and life cycle

- Humans are the only hosts of *Ancylostoma duodenale* (common hookworm) and *Necator americanus*
- Infection is common in all tropical and subtropical countries where soil is widely contaminated with human faeces and where people often walk barefoot
- 7–12 mm long adults attach themselves to the jejunal mucosa, feed by sucking blood, and mate
- Gravid females lay thousands of eggs daily that are passed on to soil with faeces

Fig. 18.1 Small bowel roundworm (contrast has entered its alimentary canal) revealed during barium meal examination.

- Infective larvae develop in a week, penetrate the skin of bare feet and migrate via the bloodstream to lungs and bronchi to pharynx, where they are swallowed and fix to the wall of the small intestine, growing to the adult form
- Egg production then takes place, completing the cycle in 6–7 weeks
- *A. duodenale* and *N. americanus* live for 6–8 years and 2–5 years, respectively
- Multiplication does not take place within the intestine.

Clinical features, diagnosis and treatment
- At the time of larval entry through the skin there may be itching and a papular dermatitis ('ground itch')
- Larval migration may cause intense eosinophilia and allergic pneumonitis (visceral larva migrans)
- Most have a low worm burden and are asymptomatic
- In heavy infection, a progressively severe iron-deficiency anaemia develops
- Lightly infected persons leaving an endemic country do not develop heavy infection later, as the worms cannot multiply within the intestine nor can the patient become reinfected from eggs in his own faeces
- Diagnosis is by demonstration of ova in stool microscopy
- Mebenadazole is the drug of choice
- Iron therapy for anaemia.

Strongyloidiasis
Epidemiology and life cycle
- Caused by *Strongyloides stercoralis*, which can exist in both a parasitic form and a free-living form in soil
- Humans are the only hosts of the parasitic form
- Found in warm humid countries of south and south-east Asia, Africa and South America
- Adult females, 2.5-mm long, live within the small intestinal mucosa and produce eggs, usually by parthenogenesis
- Eggs immediately hatch into rhabditiform larvae which can further evolve in either of two ways:

- develop into male and female rhabditiform larvae and pass out with faeces into soil, develop into adults, mate and produce eggs that hatch into infectious filariform larvae (free-living cycle), and enter another host through skin or mucous membrane contact
- directly transform into filariform larvae within intestine, and invade the host's gut wall or perianal skin (autoinfection)
- After entering the host the larvae travel to the intestines via circulation–liver–lung–bronchi–larynx–oesophagus route and mature into adults
- Autoinfection can maintain strongyloidiasis for many years after exposure to exogenous infection has ceased.

Clinical features, diagnosis and treatment
- Infections are often asymptomatic
- Urticarial or rapidly migrating serpiginious rash on buttocks, trunk or thigh (larva currans) may be caused by migrating larva
- Fluctuating eosinophilia is common
- In the immunocompromised, heavy worm load may develop from autoinfection, leading to chronic diarrhoea and malabsorption or potentially fatal dissemination to lungs and other organs
- Diagnosis is by demonstrating larvae in freshly passed stools or in duodenal aspirate or sputum (in disseminated infection, Fig. 18.2). Serology may help (not in the immunocompromised)
- Thiabendazole, albendazole and ivermectin are all effective.

Trichuriasis
Epidemiology and life cycle
- The whip-like 20–30 mm long adult parasites, *Trichuris trichura,* habitate human caecum and colon and live for up to 5 years
- Prevalent worldwide, especially in warm regions with poor sanitation
- Adult females lay eggs which are passed into soils, mature and are ingested by other hosts faecal–orally
- Released larvae mature, migrate to large intestine and begin egg-laying.

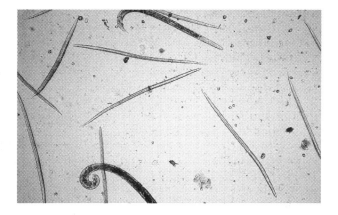

Fig. 18.2 Strongyloid larvae in sputum microscopy in an AIDS patient with disseminated strongyloidiasis.

Clinical features, diagnosis and treatment
• Most infections are asymptomatic
• Heavy infection may cause abdominal pain, blood-streaked diarrhoea (colitis) and rectal prolapse
• Diagnosis is by finding of eggs in stool specimen. Proctoscopy may reveal adult worms. Eosinophilia is rare
• Mebendazole is the drug of choice.

Tissue nematode infections

Toxocariasis

Epidemiology and clinical features
• Caused by an inflammatory reaction provoked by the larvae of animal roundworms, *Toxocara canis* (commoner) or *T. catis,* whose normal hosts are dogs and cats
• Prevalence in dogs and cats is worldwide; animals harbour adults in their intestines and excrete eggs in their faeces, widely contaminating soil
• Occurrence of toxocariasis is worldwide with a seroprevalence of asymptomatic infection of around 20% in UK and USA nursery-age children. Less common in adults
• Children usually acquire infection whilst playing on soil contaminated with puppy faeces
• Ingested eggs hatch into larvae which penetrate the gut wall and migrate via the circulation to liver, lungs, CNS and eyes, provoking eosinophilic granulomatous reaction locally

• Most infections are asymptomatic with eosinophilia only
• In heavy infections, there may be malaise, fever and weight loss, hepatomegaly, skin rashes, pneumonitis, choroidoretinitis, CNS symptoms (visceral larva migrans) or posterior retinal granulomatous mass (usually years later in older children or adults).

Diagnosis, treatment and prevention
• Diagnosis is suggested by the intense eosinophilia and hyperglobinaemia in a child with suggestive symptoms
• Eosinophilia is generally absent in ocular toxoplasmosis which tends to occur in adults
• Serology is usually confirmatory
• Stools are always negative as the larvae cannot complete life cycle in humans and mature into adults
• Antihelmintics do not generally help
• Corticosteroids may help reduce inflammatory reaction
• Prevention of contamination of soil by dog and cat faeces in children's play areas and public parks
• Prevention of pica in children
• Washing of hands after handling soil and before eating.

Cutaneous larva migrans
• This is caused when a larva of an animal hook-

worm (usually *Ancylostoma braziliensis*, which normally infects dogs and cats) penetrates human skin during walking barefoot on grounds soiled by dog or cat faeces
• Larva migrates slowly along the cutaneous lymphatics, provoking a fierce tissue reaction which shows as a slowly advancing, serpiginous inflamed track
• The larva cannot complete the life cycle as it is in the wrong host, and the condition is self-limiting within a few weeks
• Thiabendazole, albendazole or ivermectin are all effective.

Trichinellosis
• This is a disease caused by the larvae of *Trichinella* species (commonly *T. spiralis*)
• The nematode commonly infects a wide range of animals in nature; adult worms dwell within the animal intestines
• Larvae released from gravid females penetrate the intestinal wall and migrate to striated muscles and encyst
• When meat containing encysted larvae is eaten by another animal, larvae are released in the intestine, mature into adults in a few days and produce invasive larvae
• Humans eating such meat become similarly affected by larvae invading striated muscles
• Undercooked pork and wild-animal meat are the main sources of human infection
• Incidence worldwide is variable, dependent on animal rearing/feeding practices and wildlife meat eating.

Clinical features, diagnosis and treatment
• Initially there is abdominal pain and diarrhoea (during larval migration into intestinal mucosa), followed by systemic larval invasion that causes marked hypersensitivity reactions
• 1 week later muscle and joint pains, and fever develop
• Periorbital and facial oedema, and subconjunctival and subungual haemorrhages are common
• Maculopapular rash, shortness of breath and features of myocarditis may develop
• Symptoms begin to abate after 3 weeks

• After recovery, the visceral larvae are absorbed but the encysted muscle larvae may calcify after several years
• Diagnosis is suggested by the clinical features and intense eosinophilia
• Serology is positive by the 3rd week
• Muscle biopsy will show the encysted worms, often in considerable numbers
• Corticosteroids during the acute stage help, and thiabendazole is often used to kill the larvae.

Prevention
• All fresh pork and wild animal meat should be adequately cooked
• Prolonged freezing will destroy *Trichinella* cysts.

Tapeworms (cestodes)

Tapeworms are elongated ribbon-like worms, which may reach up to 25 m in length. Adults have both male and female characteristics, so a single adult worm may produce huge numbers of fertile eggs.

Life cycle
• This involves two hosts: a definitive and an intermediate.
• In the definitive host the adult worm attaches to the intestinal mucosa using suckers or grooves on its head (scolex) and gains in length by producing chains of segments (proglottids)
• Progloglottids become gravid and detach from the body and release eggs in faeces
• Eggs are swallowed by an intermediate host in whom larvae are hatched which invade across the gut wall and migrate to tissues to form cysts containing scolex
• Definitive host ingests tissues containing these cysts which dissolve in its intestine allowing the scolex to attach onto intestinal mucosa and grow into adult tapeworm
• Humans are involved as either definitive host or intermediate host.

Humans as definitive hosts
See Table 18.1.

Treatment
- Niclosamide—fears of cystcercosis from regurgitated *T. solium* eggs unfounded, though still wise to use an antiemetic
- Praziquantel—equally effective.

Prevention
- Thorough cooking of beef, pork and fish
- Inspection of meat in slaughter houses
- Freezing, prolonged refrigeration and salting will destroy scolex
- Sanitary disposal of human faeces.

Humans as intermediate hosts
See Table 18.2.

Schistosomiasis (bilharziasis)

Epidemiology and life cycle
- Caused by the blood fluke *Schistosoma* which has three major species: *S. mansoni*, *S. haematobium* and *S. japonicum*
- *Schistosoma mansoni* is endemic in Africa, the Middle East and parts of South America, *S. haematobium* is seen in Africa and the Middle East, and *S. japonicum* is prevalent in the Far East
- Humans are the principal hosts and freshwater snails of particular types are the intermediate hosts. Adult males and females live within the venules of the human bladder and intestine for years and produce eggs, which reach freshwater sources via faeces or urine
- Larvae are liberated in water and penetrate into snails, where they mature and later emerge as free-swimming larvae (cercariae). Cercariae penetrate human skin in contact with water and reach the liver via the bloodstream
- Here they mature into adults, migrate into mesenteric (*S. mansoni* and *S. japonicum*) or bladder (*S. haematobium*) veins and begin egg production (about 6–12 weeks after infection).

Pathogenesis
- Eggs are released into the lumen of the bowel or bladder and are also transported by blood to the liver, lungs and other sites

- There may be a vigorous hypersensitivity response (Katayama fever) at the time of initial worm maturation and egg deposition
- Granuloma and fibrosis form at the site of egg deposition (intestinal mucosa, urinary tract, liver, lungs, etc.)
- Chronic disease results from repeated egg deposition in large numbers and so requires repeated infections in endemic countries to produce a large worm load. Most infections are asymptomatic.

Clinical features
- *Acute schistosomiasis (Katayama fever)*: this is usually seen in visitors from non-endemic areas, 2–6 weeks after exposure—fever, urticaria, headache, abdominal pain and diarrhoea lasting for days or weeks. CNS symptoms including myelitis may develop. Eosinophilia is common
- *Chronic intestinal and hepatic schistosomiasis* (*S. mansoni* and *S. japonicum*): occasional blood-stained stools due to the formation of large bowel inflammatory polyps; possibly colorectal cancer; hepatomegaly and splenomegaly due to hepatic fibrosis and portal hypertension
- *Chronic urinary schistosomiasis* (*S. haematobium*): dysuria, frequency and haematuria due to obstructive uropathy; bladder inflammatory polyp and secondary infection; possibly bladder cancer
- Pulmonary fibrosis and hypertension
- Chronic salmonellosis (salmonellae reside within schistosome and are protected).

Diagnosis
- Demonstration of eggs in stool, urine or biopsy (rectal, bladder, liver) is required for definitive diagnosis. Remember egg deposition may take up to 12 weeks from infection
- Serology is helpful in diagnosing recent infection or acute disease in a traveller.

Treatment and prevention
- Praziquantel is the drug of choice for all three types of schistosomiasis
- Education of public in sanitary disposal of faeces and urine, measures to reduce the snail

Table 18.1 Tapeworm infections where humans are definitive hosts.

Species	Taenia saginata	Taenia solium	Diphylobothrium latum
Definitive host	Humans only, release eggs faecally	Humans only, release eggs faecally	Humans and fish-eating animals release eggs in water
Intermediate host	Cattle—ingest egg	Pigs—ingest egg	Freshwater crustacea swallow eggs (first intermediate host), in turn swallowed by fish (second intermediate host)
Human infection results from eating	Raw or undercooked beef	Raw or undercooked pork	Raw or undercooked fish
Occurrence	Sub-Saharan Africa, Middle East	Africa, south-east Asia, eastern Europe, Mexico, South America	Far East, Scandinavia, Alaska, Canada
Habitat of adult worm	Human jejunum	Human jejunum	Human or animal ileum
Scolex characteristics	4 suckers	4 suckers and hooklets	2 suckers
Characteristics of segments in stools	15–30 uterine branches	8–12 branches	Segments not passed in stools
Egg-to-egg cycle	8–12 weeks	10–14 weeks	At least 11 weeks
Clinical features	Mostly asymptomatic	Mostly asymptomatic	Mostly asymptomatic Rarely, vitamin B_{12} deficiency in massive infections due to parasitic consumption within ileum
Diagnosis	Eggs and segments in stools	Eggs and segments in stools	Eggs in stools

Table 18.2 Tapeworm infections where humans are intermediate hosts.

Features	Cysticercosis	Hydatid disease
Species	T. solium	Echinococcus granulosus Echinococcus multilocularis
Definitive host	Humans — pass eggs with faeces	Dogs — infected by eating animal flesh containing cysts, pass eggs with faeces
Intermediate host	Pigs — usual, humans, other animals	Sheep, cattle, humans, goats, horses, rodents (for E. multilocularis) infected by ingesting eggs
Occurrence	See T. solium (Table 18.1)	Australasia, Asia, South America, Africa, Middle East, eastern Europe and eastern Mediterranean, Arctic and sub-Arctic regions (E. multilocularis)
Human disease	Humans usually acquire cysticercosis by ingesting eggs via faecally contaminated food or rarely from eggs regurgitated into stomach from own intestine	Hand-to-mouth transmission of eggs in dog faeces
Pathogenesis	Hatched larvae migrate across gut wall to subcutaneous tissue, skeletal muscles and brain	Hatched larvae migrate across gut wall to liver, lungs, rarely other organs Fluid-filled unilocular hydatid cyst forms with daughter cysts developing inside Local inflammation and/or space-occupying effect — from degenerating or growing cysts Cysts expand slowly over many years Eventually calcify after dying, earlier in muscles than in brain E. multilocularis cysts are multilocular and invade host tissue by peripheral extension
Clinical features	Presentation — usually neurological, with fits, occasionally more diffuse with meningitis, encephalitis, hydrocephalous	Subcutaneous cysts may be palpable Usually silent, with incidental discovery during other investigations. May cause pressure symptoms Rupture may lead to sudden hypersensitivity reactions and danger of disseminated seeding of scolices E. multilocularis produces progressive liver or lung damage
Diagnosis	Plain X-ray of brain to show calcification CT/MR scans are better and will reveal cysts (Fig. 18.3)	Spindle-shaped calcification in skeletal muscles in plain X-ray Immunoblot serodiagnosis often helpful. MR/CT/ultrasound will show fluid-filled cyst (Fig. 18.4), often with daughter cysts Eggshell calcification in plain films is characteristic Serodiagnosis is often supportive
Treatment	Anticonvulsants	Praziquantel, albendazole Corticosteroids (suppress inflammatory response from killed parasites) Asymptomatic — usually no treatment Surgery with albendazole cover Prolonged courses of albendazole if surgery not feasible

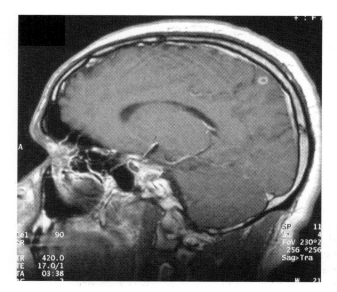

Fig. 18.3 Cerebral cysticercosis. MR scan of brain (post-gadolinium) shows a ring enhanced cyst within the brain parenchyma.

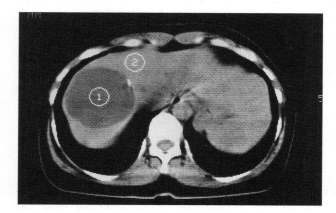

Fig. 18.4 CT scan showing a large hydatid cyst of liver (1) with some calcification (2).

population, provision of safe water for drinking and bathing, avoidance of swimming in fresh water in endemic countries.

Filariasis

This is caused by filarial worms which are nematodes and reside in the subcutaneous tissues and the lymphatics of humans. Their life cycles involve insects which carry the infective larvae. Three main clinical forms of human infections are recognized:

- lymphatic filariasis
- onchocerciasis
- loiasis.

Lymphatic filariasis
Epidemiology, life cycle and pathogenesis
- Commonly caused by *Wuchereria bancrofti*, prevalent in equatorial Africa, the Indian sub-continent, south-east Asia, and Central and South America
- Humans, the only definitive host, become infected through mosquito bites. Injected larvae migrate to the lymphatics where they mature

into adults and produce microfilariae in 6–12 months from infection
• Microfilariae migrate to peripheral blood nocturnally and are taken up by biting mosquitoes in whom they mature into infective larvae
• Adults live for many years, microfilariae for 3 months–3 years
• The tissue damage results from adult worms living within lymphatics causing inflammation, fibrosis and obstruction.

Clinical features and diagnosis
• Asymptomatic microfilaraemia is common; in others there may be episodic fever and chills, lymphangitis, lymphadenitis, epididymitis or orchitis with microfilaraemia (early disease).
• Recurrent lymphatic inflammation may lead to chronic lymphatic obstruction producing thickened, oedematous, swollen hyperkeratotic skin, involving the legs, genitalia and breasts (elephantiasis). Microfilaraemia is rare (late disease)
• Chyluria may develop from rupture of renal lymphatics
• Tropical pulmonary eosinophilia syndrome: recurrent asthma and fever, marked eosinophilia and diffuse lung infiltrates (on X-ray). Microfilaraemia is absent
• The early manifestations are usually diagnosed by demonstrating microfilaria in blood film
• Chronic filariasis is a clinical diagnosis but filarial antibodies are usually present in blood
• Tropical pulmonary eosinophilia is diagnosed by demonstrating high-titre serum filarial antibodies in association with characteristic clinical, haematological and radiological findings.

Treatment and prevention
• Diethylcarbamazepine rapidly eliminates microfilaria and symptoms of early disease. Adult worms are killed less easily and repeat courses are necessary
• Febrile and local tissue reactions are common during treatment due to disintegrating parasites, and corticosteroids are helpful
• Obstructive features of chronic disease may be helped by surgical reconstruction

• Vector control, insect repellents and use of mosquito nets and screens.

Onchocerciasis
Epidemiology and life cycle
• Caused by *Onchocerca volvulus*, the disease is endemic in equatorial Africa and South America and is a leading cause of blindness
• Humans only are affected, through inoculation of larvae by a biting female blackfly
• Within the skin the larvae mature into adults after a few months. Several males and females lie coiled together in round bundles in subcutaneous tissue and produce microfilariae
• These migrate to skin and eyes. The female blackfly ingests microfilariae while biting the skin.

Clinical features and diagnosis
• Dermatitis: most common (later, the skin becomes thickened and wrinkled)
• Visible or palpable subcutaneous nodules, enlarged lymph glands and visual impairment due to conjunctivitis, keratitis, anterior uveitis and choroidoretinitis
• Diagnosis is by demonstrating microfilariae in skin biopsy or cornea (slit lamp examination), or adult worms in excised subcutaneous nodules.

Treatment and prevention
• Ivermectin as a single dose repeated yearly (until adult worms die) is the drug of choice
• Avoidance of bites, vector control measures and treatment of infection to reduce the reservoir of infection.

Loiasis
Epidemiology and pathogenesis
• Caused by adult *Loa loa* worms, and seen mainly in central African rain forests
• Vectors are horsefly or deerfly of the genus *Chrysops*
• Larvae injected into humans during the bite of an infected fly mature into adults, which live in subcutaneous tissues and produce microfilariae that migrate to peripheral blood and are taken up by biting vectors.

Clinical features

- Asymptomatic microfilaraemia is common
- Transient subcutaneous swellings (Calabar swelling, from migrating worms), commonly on extremities
- Worms may migrate across the scleral or conjunctival tissue
- Fever and urticaria may occur.

Diagnosis

- Detection of microfilaria in blood or of adult worm in subcutaneous biopsy or from eye
- Suggestive clinical picture; eosinophilia and positive serology is acceptable in returning travellers.

Treatment and prevention

- Diethylcarbamazepine. Repeated courses may be necessary. There may be severe systemic reactions; corticosteroids and inpatient supervision are recommended
- Insect-repellent measures, prophylactic diethylcarbamzine.

Index